INTRODUCTORY ALGEBRA
CHAPTER 8, TEST B, PAGE 2

4. Combine the following fractions and simplify.

   (a) $\dfrac{5x}{x+1} + \dfrac{2x}{x+1}$

   (b) $\dfrac{3}{4x} - \dfrac{11}{4x}$

   (c) $\dfrac{4x+1}{5} - \dfrac{x-7}{5}$

   4(a) _____

   (b) _____

   (c) _____

5. Combine the fractions as indicated and reduce the results to lowest terms.

   (a) $\dfrac{1}{3x} + \dfrac{3}{4x}$

   (b) $\dfrac{x+1}{x} - \dfrac{y-2}{y}$

   (c) $\dfrac{6}{x} - \dfrac{2}{x-2}$

   (d) $3 + \dfrac{2}{x+1}$

   5(a) _____

   (b) _____

   (c) _____

   (d) _____

6. Simplify each of the following complex fractions.

   (a) $\dfrac{2 + \dfrac{3}{x}}{x}$

   (b) $\dfrac{\dfrac{x}{3}}{\dfrac{5x}{3}}$

   6(a) _____

   (b) _____

7. Solve the following equations and check.

   (a) $\dfrac{2x}{3} = 4$

   (b) $\dfrac{5x}{x+1} = 4$

   7(a) _____

   (b) _____

8. Using a power blower, Ron's lawn can be cleared of leaves in two hours. Raking his lawn using a hand rake, however, requires eight hours. How long would it take to clear Ron's lawn of leaves if both a power blower and a hand rake were used?

   8. _____

INTRODUCTORY ALGEBRA
CHAPTER 8, TEST B, PAGE 1

1. Reduce the following fractions to lowest terms.

   (a) $\dfrac{30x^2y}{30xy}$ (b) $\dfrac{5x+5y}{10x+10y}$

   (c) $\dfrac{x^2-y^2}{(x-y)^2}$ (d) $\dfrac{x^2-16}{x^2-7x+12}$

   (e) $\dfrac{1-x}{x^2-1}$

   1(a) _____

   (b) _____

   (c) _____

   (d) _____

   (e) _____

2. Multiply the following fractions and express the product in reduced form.

   (a) $\dfrac{7ab}{5cd} \cdot \dfrac{15c^2d^2}{28a^2b^2}$ (b) $\dfrac{3}{2x} \cdot 22x^2$

   (c) $\dfrac{8x-8y}{15x^2y^2} \cdot \dfrac{21x^3y^3}{7(x-y)^2}$ (d) $\dfrac{a^2-1}{15-15a} \cdot \dfrac{6-2a}{6a+6}$

   2(a) _____

   (b) _____

   (c) _____

   (d) _____

3. Divide the following fractions. Reduce the quotient to lowest terms.

   (a) $\dfrac{2x^4y}{5} \div \dfrac{4x}{25y}$ (b) $8xy \div \dfrac{24y}{x}$

   (c) $\dfrac{3x-6}{4x+4} \div \dfrac{x-2}{(x+1)^2}$ (d) $\dfrac{x^2-3x+2}{x^2-1} \div \dfrac{2-x}{1-x}$

   3(a) _____

   (b) _____

   (c) _____

   (d) _____

101

INTRODUCTORY ALGEBRA
CHAPTER 9, TEST C, PAGE 1

1. Perform the indicated operation.

   (a) $\sqrt{4x^6}$ (b) $\sqrt{25x^6y^8}$

2. Simplify the following radicals.

   (a) $\sqrt{90}$ (b) $\frac{1}{2}\sqrt{48}$

   (c) $\sqrt{x^4y^5}$ (d) $-3\sqrt{20x^2y}$

3. Simplify the following radicals.

   (a) $5\sqrt{\frac{16}{25}}$ (b) $\sqrt{\frac{54}{49}}$

   (c) $6\sqrt{\frac{28}{16}}$

4. Combine the following radicals.

   (a) $6\sqrt{7} - 9\sqrt{7}$

   (b) $2\sqrt{7} + 2\sqrt{6} - 3\sqrt{7} + \sqrt{6}$

   (c) $2\sqrt{27} + \frac{1}{2}\sqrt{12}$

1(a)_____

(b)_____

2(a)_____

(b)_____

(c)_____

(d)_____

3(a)_____

(b)_____

(c)_____

4(a)_____

(b)_____

(c)_____

Instructor's Manual to Accompany

INTRODUCTORY ALGEBRA

Michael A. Gallo
Monroe Community College

Charles F. Kiehl
SUNY, Brockport

WEST PUBLISHING COMPANY
St. Paul   New York   Los Angeles   San Francisco

INTRODUCTORY ALGEBRA
CHAPTER 9, TEST C, PAGE 2

5. Multiply the following radicals and express the product in its simplest form.

   (a) $\sqrt{4} \cdot \sqrt{9}$

   (b) $3\sqrt{x} \cdot \sqrt{49x}$

   (c) $(-2\sqrt{x})^2$

   (d) $\sqrt{3}(\sqrt{2} - \sqrt{7})$

5(a)_____

(b)_____

(c)_____

(d)_____

6. Express each quotient in its simplest radical form.

   (a) $\dfrac{\sqrt{30}}{3\sqrt{15}}$

   (b) $\dfrac{\sqrt{2}}{\sqrt{5}}$

   (c) $\dfrac{6}{\sqrt{12}}$

   (d) $16\sqrt{\dfrac{3}{18}}$

6(a)_____

(b)_____

(c)_____

(d)_____

7. Solve and check the following radical equations.

   (a) $\sqrt{3x} = \dfrac{1}{3}$

   (b) $\sqrt{3x} - 7 = -1$

7(a)_____

(b)_____

INTRODUCTORY ALGEBRA
CHAPTER 9, TEST B, PAGE 2

5. Multiply the following radicals and express the product in its simplest form.

   (a) $\sqrt{5} \cdot \sqrt{5}$

   (b) $2 \cdot 5\sqrt{10}$

   (c) $(2x\sqrt{7x})^2$

   (d) $5\sqrt{10}\,(2\sqrt{5} - \sqrt{15})$

5(a) _____
(b) _____
(c) _____
(d) _____

6. Express each quotient in its simplest radical form.

   (a) $\dfrac{6\sqrt{30}}{3\sqrt{5}}$

   (b) $\dfrac{1}{\sqrt{3}}$

   (c) $\dfrac{7\sqrt{3}}{4\sqrt{3}}$

   (d) $3\sqrt{\dfrac{5}{12}}$

6(a) _____
(b) _____
(c) _____
(d) _____

7. Solve and check the following radical equations.

   (a) $3\sqrt{x} = 9$

   (b) $\sqrt{3x} + 1 = 10$

7(a) _____
(b) _____

INTRODUCTORY ALGEBRA
CHAPTER 9, TEST B, PAGE 1

1. Perform the indicated operation.

   (a) $\sqrt{y^6}$ (b) $\sqrt{144x^4y^2}$

   1(a) _____

   (b) _____

2. Simplify the following radicals.

   (a) $\sqrt{20}$ (b) $-3\sqrt{24}$

   (c) $\sqrt{x^5y^5}$ (d) $2\sqrt{12x^3}$

   2(a) _____

   (b) _____

   (c) _____

   (d) _____

3. Simplify the following radicals.

   (a) $\sqrt{\dfrac{1}{9}}$ (b) $\sqrt{\dfrac{18}{16}}$

   (c) $9\sqrt{\dfrac{24}{81}}$

   3(a) _____

   (b) _____

   (c) _____

4. Combine the following radicals.

   (a) $4\sqrt{5} - \sqrt{5}$

   (b) $\sqrt{2} - \sqrt{3} + 4\sqrt{2} - 5\sqrt{3}$

   (c) $6\sqrt{2} - 2\sqrt{8}$

   4(a) _____

   (b) _____

   (c) _____

113

INTRODUCTORY ALGEBRA
CHAPTER 8, TEST D, PAGE 2

4. Combine the following fractions and simplify.

   (a) $\dfrac{3x-1}{x+2} + \dfrac{x+1}{x+2}$ (b) $\dfrac{x^2}{x-4} - \dfrac{16}{x-4}$

   (c) $\dfrac{12x}{5y} - \dfrac{2x+5}{5y}$

   4(a) _____
   (b) _____
   (c) _____

5. Combine the fractions as indicated and reduce the results to lowest terms.

   (a) $\dfrac{3}{a^2} - \dfrac{2}{a}$ (b) $\dfrac{2}{a+1} + \dfrac{2}{2a-1}$

   (c) $\dfrac{x}{9x^2-1} + \dfrac{1}{6x+2}$ (d) $x - \dfrac{3x}{4}$

   5(a) _____
   (b) _____
   (c) _____
   (d) _____

6. Simplify each of the following complex fractions.

   (a) $\dfrac{\dfrac{x}{y}}{\dfrac{x^2}{y^2}}$ (b) $\dfrac{4 - \dfrac{2}{xy}}{1 - \dfrac{1}{xy}}$

   6(a) _____
   (b) _____

7. Solve the following equations and check.

   (a) $\dfrac{x-5}{3} = 2$ (b) $\dfrac{x-2}{x+1} = \dfrac{x+1}{x-2}$

   7(a) _____
   (b) _____

8. It takes seven minutes to complete a certain job by hand and only three minutes if a power tool is used. How long would it take if both were used?

   8. _____

INTRODUCTORY ALGEBRA
CHAPTER 8, TEST D, PAGE 1

1. Reduce the following fractions to lowest terms.

(a) $\dfrac{-16x^2y^3z}{-2xy^3z^2}$ (b) $\dfrac{2x+3}{10x+15}$ 1(a) _____

(b) _____

(c) $\dfrac{4x+4y}{x^2-y^2}$ (d) $\dfrac{3x-9}{x^2-5x+6}$ (c) _____

(d) _____

(e) $\dfrac{x^2-6x+9}{9-x^2}$ (e) _____

2. Multiply the following fractions and express the product in reduced form.

(a) $\dfrac{20a}{b^2} \cdot \dfrac{b}{28a^3}$ (b) $4\pi r^2 \cdot \dfrac{1}{2\pi r}$ 2(a) _____

(b) _____

(c) $\dfrac{x^2+x}{x^2} \cdot \dfrac{3x-3}{x^2-1}$ (d) $\dfrac{x^2-y^2}{4} \cdot \dfrac{3x^2+3y^2}{x^4-y^4}$ (c) _____

(d) _____

3. Divide the following fractions. Reduce the quotient to lowest terms.

(a) $\dfrac{3y}{4x^2} \div \dfrac{16x^3}{15y}$ (b) $36x^2 \div \dfrac{15x^3}{2y^2}$ 3(a) _____

(b) _____

(c) $\dfrac{x^2+2x+1}{x-3} \div \dfrac{x+1}{2}$ (d) $\dfrac{x^2-100}{8x-64} \div \dfrac{2x+20}{x^3-64x}$ (c) _____

(d) _____

COPYRIGHT © 1984 by WEST PUBLISHING CO.
50 West Kellogg Boulevard
P.O. Box 43526
St. Paul, MN 55164

All rights reserved
Printed in the United States of America
ISBN 0-314-78002-5
1st Reprint—1985

Included in this Instructor's Manual are six different but parallel chapter tests for each chapter in the text book, two final examinations, answers to the chapter tests included in this manual, answers to the two final examinations included in this manual, and answers to all exercises found in the text book.

TABLE OF CONTENTS

Chapter 1, Tests A, B, C, D, E, F 3

Chapter 2, Tests A, B, C, D, E, F 27

Chapter 3, Tests A, B, C, D, E, F 39

Chapter 4, Tests A, B, C, D, E, F 51

Chapter 5, Tests A, B, C, D, E, F 63

Chapter 6, Tests A, B, C, D, E, F 75

Chapter 7, Tests A, B, C, D, E, F 87

Chapter 8, Tests A, B, C, D, E, F 99

Chapter 9, Tests A, B, C, D, E, F 111

Chapter 10, Tests A, B, C, D, E, F 123

Chapter 11, Tests A, B, C, D, E, F 147

Chapter 12, Tests A, B, C, D, E, F 159

Final Examinations, Forms A and B 171

Answers to Chapter Tests 185

Answers to Final **Examinations** **211**

Answers to Chapter Exercises 215

CHAPTER TESTS

INTRODUCTORY ALGEBRA
CHAPTER 7, TEST B, PAGE 2

4. Factor the following trinomials.

(a) $2x^2 + 7x + 6$ 4(a) _____

(b) $3x^2 - 5x + 2$ (b) _____

(c) $4x^2 + 2x - 6$ (c) _____

(d) $20x^2 - 19x + 3$ (d) _____

5. Factor the difference of two squares.

(a) $z^2 - 100$ 5(a) _____

(b) $9x^2 - 49$ (b) _____

6. Factor completely the following expressions.

(a) $7x^2 - 14x - 21$ 6(a) _____

(b) $x^4 - 1$ (b) _____

(c) $3ax^2 - 9ax + 6a$ (c) _____

7. Solve and check for the roots in the following equations.

(a) $x^2 - 5x = 0$ 7(a) _____

(b) $2x^2 - 50 = 0$ (b) _____

(c) $x^2 + 5x - 14 = 0$ (c) _____

(d) $3x^2 + 13x + 14 = 0$ (d) _____

INTRODUCTORY ALGEBRA
CHAPTER 7, TEST B, PAGE 1

1. Factor each of the following polynomials whose terms have a common factor.

(a) $3a^2 + 9$ 1(a) _____

(b) $-3y - 6y^2$ (b) _____

(c) $x(x+1) + 2(x+1)$ (c) _____

2. Find each of the indicated products by the foil method.

(a) $(x+1)(x+1)$ 2(a) _____

(b) $(2x-5)(x-1)$ (b) _____

(c) $(5y-3)(y+2)$ (c) _____

(d) $(2x-3)^2$ (d) _____

3. Factor the following trinomials.

(a) $x^2 + 5x + 4$ 3(a) _____

(b) $x^2 - 7x - 8$ (b) _____

(c) $x^2 + x - 6$ (c) _____

(d) $x^2 - 10x + 9$ (d) _____

INTRODUCTORY ALGEBRA
CHAPTER 6, TEST B, PAGE 2

6. Find the product of the following.

(a) $(-3x^5)(3x)$ (b) $-4x(-5x^2-4x+8)$

(c) $(x+7)(x+2)$ (d) $(x+1)(1+2x-2x^2)$

6(a) _____

(b) _____

(c) _____

(d) _____

7. Divide the following.

(a) $\dfrac{y^5}{y}$ (b) $\dfrac{x^2y^3}{x^2y^2}$

(c) $\dfrac{y}{y^6}$ (d) $\dfrac{x}{xy^2}$

7(a) _____

(b) _____

(c) _____

(d) _____

8. Divide the following.

(a) $(-5x^2y) \div (-15x^2y^4)$

(b) $(x+y+z) \div (xyz)$

(c) $(3x^2+19x+20) \div (3x+4)$

8(a) _____

(b) _____

(c) _____

78

INTRODUCTORY ALGEBRA
CHAPTER 6, TEST B, PAGE 1

1. Add the following monomials.

   (a) $7x + (-13x)$ (b) $4x^2 + (-4x^2)$

   1(a) _____

   (b) _____

   (c) $3(a+b)$
       $-4(a+b)$
       $-2(a+b)$

   (c) _____

2. Add the following polynomials.

   (a) $2x - y$
       $x + 6y$

   (b) $(3-3x) + (3+x)$

   (c) $(3x^2+xy+2y^2) + (3y^2-xy+5x^2) + (x^2-y^2)$

   2(a) _____

   (b) _____

   (c) _____

3. Subtract the following monomials.

   (a) $-6x$
       $-(2x)$

   (b) $12x^2 - (-x^2)$

   3(a) _____

   (b) _____

4. Subtract the following polynomials.

   (a) $4ab+2a-3$
       $-(-ab-2a-5)$

   (b) $3x - (2x+1)$

   (c) $(12x+15y) - (7x-3y)$

   4(a) _____

   (b) _____

   (c) _____

5. Evaluate the following expressions.

   (a) $a \cdot a^4$ (b) $(y^2)^4$

   (c) $(xy^2)^2$ (d) $(3xy)^3$

   5(a) _____

   (b) _____

   (c) _____

   (d) _____

77

INTRODUCTORY ALGEBRA
CHAPTER 1, TEST B, PAGE 1

1. State the coordinate of point A.   1._____

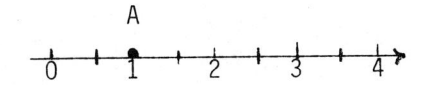

   A
   ⊢—•—⊢—⊢—⊢—⊢—⊢—⊢—→
   0   1   2   3   4

2. Express the decimal 0.05 as a fraction.   2._____

3. Express the fraction $\frac{1}{4}$ as a decimal.   3._____

4. Express the mixed number $2\frac{2}{3}$ 4._____
   as an improper fraction.

5. Express the improper fraction $\frac{9}{4}$ 5._____
   as a mixed number.

6. Compare the following numbers.
   (Use the symbols <, > or = .)

   (a) 7, 0   6(a)_____

   (b) 3, 3   (b)_____

   (c) 2, 6   (c)_____

7. Compare the following decimals.
   (Use the symbols <, > or = .)

   (a) 0.12, 0.06   7(a)_____

   (b) 2.99, 5.1   (b)_____

   (c) 0.125, 0.125   (c)_____

INTRODUCTORY ALGEBRA
CHAPTER 1, TEST B, PAGE 2

8. Compare the following fractions.
   (Use the symbols <, > or =.)

   (a) $\frac{1}{2}, \frac{3}{4}$ 8(a) _____

   (b) $\frac{3}{8}, \frac{9}{24}$ (b) _____

   (c) $\frac{7}{9}, \frac{3}{4}$ (c) _____

9. Add the following numbers.   9(a) _____

   (a) 521 + 235   (b) 204 + 319   (b) _____

10. Add the following decimals.   10(a) _____

    (a) 0.147 + 0.032   (b) 1.35 + 0.071   (b) _____

11. Add the following fractions.   11(a) _____
    (Reduce answers to lowest terms.)

    (a) $\frac{1}{2} + \frac{3}{5}$ (b) $\frac{1}{3} + \frac{2}{3}$ (b) _____

12. Add the following mixed numbers.   12(a) _____
    (Reduce answers to lowest terms.)

    (a) $2\frac{1}{4} + 3\frac{3}{4}$ (b) $1\frac{1}{2} + 2\frac{1}{3}$ (b) _____

13. Subtract the following numbers.   13(a) _____

    (a) 1628 - 204   (b) 100 - 72   (b) _____

INTRODUCTORY ALGEBRA
CHAPTER 1, TEST B, PAGE 3

14. Subtract the following als. 14(a)_____

    (a) 19.3 - 10   ) 10.5 - 0.75   (b)_____

15. Subtract the follow ctions.   15(a)_____
    (Reduce answers to terms.)

    (a) $\frac{7}{2} - \frac{1}{2}$ (b) $\frac{4}{5} - \frac{1}{2}$ (b)_____

16. Subtract the fo mixed numbers.   16(a)_____
    (Reduce answers west terms.)

    (a) $2\frac{3}{4} - 1\frac{2}{4}$ (b) $12 - \frac{3}{5}$ (b)_____

17. Multiply the wing numbers.   17(a)_____

    (a) 299 x   (b) 450 x 41   (b)_____

18. Multiply llowing decimals.   18(a)_____

    (a) 0.   5   (b) 0.04 x 1.3   (b)_____

19. Multi following fractions.   19(a)_____
    (Red vers to lowest terms.)

    (a) $\frac{}{3}$ (b) $\frac{6}{5} \times \frac{5}{18}$ (b)_____

20. M the following mixed numbers.   20(a)_____
    answers to lowest terms.)

    $\frac{}{3} \times 4\frac{1}{2}$ (b) $6 \times \frac{2}{3}$ (b)_____

INTRODUCTORY ALGEBRA
CHAPTER 1, TEST B, PAGE 4

21. Divide the following numbers.

    (a) $80 \div 5$ (b) $1330 \div 19$

21(a) _____

(b) _____

22. Divide the following decimals.

    (a) $3.22 \div 14$ (b) $0.0276 \div 0.12$

22(a) _____

(b) _____

23. Divide the following fractions.
(Reduce answers to lowest terms.)

    (a) $\frac{3}{4} \div \frac{9}{16}$ (b) $\frac{20}{49} \div \frac{5}{7}$

23(a) _____

(b) _____

24. Divide the following mixed numbers.
(Reduce answers to lowest terms.)

    (a) $6\frac{7}{8} \div 3\frac{5}{12}$ (b) $\frac{4}{7} \div 18$

24(a) _____

(b) _____

25. Express the following product using exponents.

    (a) $x \cdot x \cdot 3 \cdot 3$

    (b) $4 \cdot 4 \cdot 4 \cdot 4 \cdot 4$

25(a) _____

(b) _____

26. Evaluate.

    (a) $5^2 - (2^3)$ (b) $3^2 \cdot 2^2$

26(a) _____

(b) _____

27. Evaluate using the proper order of operations.

    (a) $5 + 10 \div 5$ (b) $4 \cdot (7 - 2)$

27(a) _____

(b) _____

INTRODUCTORY ALGEBRA
CHAPTER 1, TEST C, PAGE 1

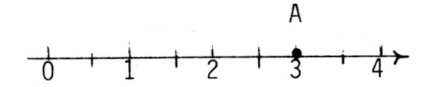

1. State the coordinate of point A. 1._____

                A
   +---+---+---+---+---+---+---+---+--->
   0   1   2   3   4

2. Express the decimal 0.25 as a fraction. 2._____

3. Express the fraction $\frac{1}{2}$ as a decimal. 3._____

4. Express the mixed number $3\frac{5}{8}$ as an improper fraction. 4._____

5. Express the improper fraction $\frac{3}{2}$ as a mixed number. 5._____

6. Compare the following numbers.
   (Use the symbols $<$, $>$ or $=$.)

   (a) 4, 4   6(a)_____

   (b) 1, 9   (b)_____

   (c) 5, 2   (c)_____

7. Compare the following decimals.
   (Use the symbols $<$, $>$ or $=$.)

   (a) 0, 0.001   7(a)_____

   (b) 3.12, 3.1200   (b)_____

   (c) 0.4, 0.125   (c)_____

INTRODUCTORY ALGEBRA
CHAPTER 1, TEST C, PAGE 2

8. Compare the following fractions.
   (Use the symbols <, > or =.)

   (a) $\frac{6}{9}, \frac{2}{3}$ 8(a)_____

   (b) $\frac{1}{8}, \frac{3}{8}$ (b)_____

   (c) $2\frac{1}{3}, \frac{2}{3}$ (c)_____

9. Add the following numbers.   9(a)_____

   (a) 388 + 99   (b) 2481 + 5417   (b)_____

10. Add the following decimals.   10(a)_____

    (a) 2.99 + 0.009   (b) 2.95 + 3.05   (b)_____

11. Add the following fractions.   11(a)_____
    (Reduce answers to lowest terms.)

    (a) $\frac{2}{5} + \frac{1}{5}$ (b) $\frac{1}{2} + \frac{3}{4}$ (b)_____

12. Add the following mixed numbers.   12(a)_____
    (Reduce answers to lowest terms.)

    (a) $6\frac{5}{8} + 7\frac{5}{8}$ (b) $4\frac{5}{6} + 1\frac{1}{4}$ (b)_____

13. Subtract the following numbers.   13(a)_____

    (a) 1492 - 1312   (b) 4038 - 162   (b)_____

INTRODUCTORY ALGEBRA
CHAPTER 1, TEST C, PAGE 3

14. Subtract the following decimals. 14(a)_____

    (a) 14.7 - 9.9   (b) 1.25 - 0.125   (b)_____

15. Subtract the following fractions. 15(a)_____
    (Reduce answers to lowest terms.)

    (a) $\frac{7}{9} - \frac{4}{9}$ (b) $\frac{7}{8} - \frac{3}{4}$ (b)_____

16. Subtract the following mixed numbers. 16(a)_____
    (Reduce answers to lowest terms.)

    (a) $9\frac{2}{5} - 6\frac{4}{5}$ (b) $6\frac{7}{12} - 3\frac{1}{3}$ (b)_____

17. Multiply the following numbers. 17(a)_____

    (a) 106 x 7   (b) 109 x 55   (b)_____

18. Multiply the following decimals. 18(a)_____

    (a) 1.7 x 0.25   (b) 2.5 x 3.1   (b)_____

19. Multiply the following fractions. 19(a)_____
    (Reduce answers to lowest terms.)

    (a) $\frac{1}{3} \times \frac{1}{4}$ (b) $\frac{3}{5} \times \frac{10}{9}$ (b)_____

20. Multiply the following mixed numbers. 20(a)_____
    (Reduce answers to lowest terms.)

    (a) $8\frac{3}{4} \times \frac{7}{8}$ (b) $2\frac{2}{3} \times 3\frac{3}{5}$ (b)_____

INTRODUCTORY ALGEBRA
CHAPTER 1, TEST C, PAGE 4

21. Divide the following numbers. 21(a)_____

    (a) 320 ÷ 8   (b) 496 ÷ 31   (b)_____

22. Divide the following decimals. 22(a)_____

    (a) 48.18 ÷ 6   (b) 2.04 ÷ 0.04   (b)_____

23. Divide the following fractions. 23(a)_____
(Reduce answers to lowest terms.)

    (a) $\frac{4}{5} \div \frac{3}{10}$ (b) $\frac{5}{9} \div \frac{25}{27}$ (b)_____

24. Divide the following mixed numbers. 24(a)_____
(Reduce answers to lowest terms.)

    (a) $2\frac{1}{2} \div 3\frac{3}{8}$ (b) $9 \div \frac{3}{8}$ (b)_____

25. Express the following product using exponents. 25(a)_____

    (a) 7 · 7 · 7 · x · x

    (b) y · y · y · 2   (b)_____

26. Evaluate. 26(a)_____

    (a) $2^2 + 2$ (b) $10^2 \cdot 2$ (b)_____

27. Evaluate using the proper order of operations. 27(a)_____

    (a) 4 x 2 + 3 x 2   (b) 15 ÷ (5 - 2)

    (b)_____

INTRODUCTORY ALGEBRA
CHAPTER 1, TEST D, PAGE 1

1. State the coordinate of point A. 1. _____

```
   A
   •────┼────┼────┼────┼──►
   0    1    2    3    4
```

2. Express the decimal 0.7 as a fraction. 2. _____

3. Express the fraction $\frac{1}{8}$ as a decimal. 3. _____

4. Express the mixed number $3\frac{1}{4}$ as an improper fraction. 4. _____

5. Express the improper fraction $\frac{4}{4}$ as a mixed number. 5. _____

6. Compare the following numbers.
   (Use the symbols $<$, $>$ or $=$.)

   (a) 7, 6 6(a) _____

   (b) 0, 9 (b) _____

   (c) 2, 2 (c) _____

7. Compare the following decimals.
   (Use the symbols $<$, $>$ or $=$.)

   (a) 0.4, 0.40 7(a) _____

   (b) 2.1, 0.9 (b) _____

   (c) 0.135, 0.8 (c) _____

INTRODUCTORY ALGEBRA
CHAPTER 1, TEST D, PAGE 2

8. Compare the following fractions.
   (Use the symbols <, > or = .)

   (a) $2\frac{1}{2}$, $2\frac{1}{4}$ 8(a)_____

   (b) $\frac{1}{2}$, $\frac{4}{8}$ (b)_____

   (c) $\frac{2}{3}$, $\frac{3}{4}$ (c)_____

9. Add the following numbers.   9(a)_____

   (a) 306 + 472   (b) 2007 + 108   (b)_____

10. Add the following decimals.   10(a)_____

    (a) 0.32 + 0.222   (b) 3.208 + 1.003   (b)_____

11. Add the following fractions.   11(a)_____
    (Reduce answers to lowest terms.)

    (a) $\frac{1}{3} + \frac{5}{6}$ (b) $\frac{1}{8} + \frac{3}{8}$ (b)_____

12. Add the following mixed numbers.   12(a)_____
    (Reduce answers to lowest terms.)

    (a) $2\frac{1}{2} + 3\frac{1}{2}$ (b) $2\frac{1}{2} + 3\frac{3}{4}$ (b)_____

13. Subtract the following numbers.   13(a)_____

    (a) 270 - 130   (b) 3004 - 28   (b)_____

INTRODUCTORY ALGEBRA
CHAPTER 1, TEST D, PAGE 3

14. Subtract the following decimals.

    (a) 13.05 - 7.3   (b) 16.05 - 4.89

14(a)_____
(b)_____

15. Subtract the following fractions. (Reduce answers to lowest terms.)

    (a) $\frac{3}{4} - \frac{1}{4}$ (b) $\frac{2}{3} - \frac{1}{2}$

15(a)_____
(b)_____

16. Subtract the following mixed numbers. (Reduce answers to lowest terms.)

    (a) $9\frac{7}{15} - \frac{12}{15}$ (b) $6\frac{3}{4} - 2\frac{5}{6}$

16(a)_____
(b)_____

17. Multiply the following numbers.

    (a) 365 x 8   (b) 386 x 30

17(a)_____
(b)_____

18. Multiply the following decimals.

    (a) 2.9 x 0.007   (b) 0.02 x 0.03

18(a)_____
(b)_____

19. Multiply the following fractions. (Reduce answers to lowest terms.)

    (a) $\frac{5}{6} \times \frac{5}{7}$ (b) $\frac{4}{6} \times \frac{9}{18}$

19(a)_____
(b)_____

20. Multiply the following mixed numbers. (Reduce answers to lowest terms.)

    (a) $3\frac{1}{8} \times 6\frac{2}{5}$ (b) $\frac{3}{5} \times 10$

20(a)_____
(b)_____

INTRODUCTORY ALGEBRA
CHAPTER 1, TEST D, PAGE 4

21. Divide the following numbers.

    (a) $312 \div 3$ (b) $1505 \div 35$

21(a)_____

(b)_____

22. Divide the following decimals.

    (a) $172.2 \div 21$ (b) $44.2 \div 1.7$

22(a)_____

(b)_____

23. Divide the following fractions.
    (Reduce answers to lowest terms.)

    (a) $\frac{2}{3} \div \frac{3}{10}$ (b) $\frac{4}{9} \div \frac{6}{15}$

23(a)_____

(b)_____

24. Divide the following mixed numbers.
    (Reduce answers to lowest terms.)

    (a) $2\frac{1}{4} \div \frac{4}{3}$ (b) $10 \div \frac{5}{9}$

24(a)_____

(b)_____

25. Express the following product using exponents.

    (a) $x \cdot x \cdot 2$

    (b) $2 \cdot x \cdot 2 \cdot y \cdot 2 \cdot x$

25(a)_____

(b)_____

26. Evaluate.

    (a) $3^2 + 3^2$ (b) $2^3 \cdot 3^2$

26(a)_____

(b)_____

27. Evaluate using the proper order of operations.

    (a) $4 + 9 \div 3$ (b) $6 \cdot (3-2) \cdot (4 \div 2)$

27(a)_____

(b)_____

INTRODUCTORY ALGEBRA
CHAPTER 1, TEST E, PAGE 1

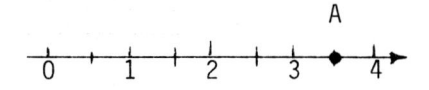

1. State the coordinate of point A. 1._____

```
            A
 +--+--+--+--●--+→
 0  1  2  3     4
```

2. Express the decimal 0.50 as a fraction. 2._____

3. Express the fraction $\frac{3}{4}$ as a decimal. 3._____

4. Express the mixed number $2\frac{7}{8}$ as an improper fraction. 4._____

5. Express the improper fraction $\frac{11}{5}$ as a mixed number. 5._____

6. Compare the following numbers.
   (Use the symbols $<$, $>$ or $=$.)

   (a) 2, 4   6(a)_____

   (b) 5, 5   (b)_____

   (c) 1, 0   (c)_____

7. Compare the following decimals.
   (Use the symbols $<$, $>$ or $=$.)

   (a) 0.9, 0.89   7(a)_____

   (b) 0.09, 0.9   (b)_____

   (c) 3.001, 3.001   (c)_____

INTRODUCTORY ALGEBRA
CHAPTER 1, TEST E, PAGE 2

8. Compare the following fractions.
   (Use the symbols $<$, $>$ or $=$.)

   (a) $1, \frac{10}{9}$ 8(a) _____

   (b) $\frac{5}{7}, \frac{6}{8}$ (b) _____

   (c) $\frac{1}{3}, \frac{3}{9}$ (c) _____

9. Add the following numbers.   9(a) _____

   (a) $314 + 39$ (b) $1234 + 4321$ (b) _____

10. Add the following decimals.   10(a) _____

    (a) $0.45 + 1.045$ (b) $2.74 + 0.76$ (b) _____

11. Add the following fractions.   11(a) _____
    (Reduce answers to lowest terms.)

    (a) $\frac{1}{4} + \frac{1}{4}$ (b) $\frac{1}{3} + \frac{2}{5}$ (b) _____

12. Add the following mixed numbers.   12(a) _____
    (Reduce answers to lowest terms.)

    (a) $3\frac{1}{5} + 2\frac{2}{5}$ (b) $1\frac{2}{5} + 3\frac{1}{2}$ (b) _____

13. Subtract the following numbers.   13(a) _____

    (a) $2471 - 1060$ (b) $3124 - 1895$ (b) _____

INTRODUCTORY ALGEBRA
CHAPTER 1, TEST E, PAGE 3

14. Subtract the following decimals.

    (a) 0.023 - 0.01   (b) 13 - 0.13

14(a)_____

(b)_____

15. Subtract the following fractions.
(Reduce answers to lowest terms.)

    (a) $\frac{3}{5} - \frac{1}{5}$ (b) $\frac{2}{3} - \frac{2}{5}$

15(a)_____

(b)_____

16. Subtract the following mixed numbers.
(Reduce answers to lowest terms.)

    (a) $6\frac{4}{5} - \frac{1}{5}$ (b) $4 - 2\frac{4}{5}$

16(a)_____

(b)_____

17. Multiply the following numbers.

    (a) 720 x 6   (b) 709 x 46

17(a)_____

(b)_____

18. Multiply the following decimals.

    (a) 2.34 x 24.5   (b) 0.18 x 7

18(a)_____

(b)_____

19. Multiply the following fractions.
(Reduce answers to lowest terms.)

    (a) $\frac{1}{4} \times \frac{3}{5}$ (b) $\frac{4}{9} \times \frac{3}{12}$

19(a)_____

(b)_____

20. Multiply the following mixed numbers.
(Reduce answers to lowest terms.)

    (a) $\frac{5}{6} \times 2\frac{1}{3}$ (b) $1\frac{1}{2} \times 2\frac{4}{7}$

20(a)_____

(b)_____

INTRODUCTORY ALGEBRA
CHAPTER 1, TEST E, PAGE 4

21. Divide the following numbers.

   (a) 364 ÷ 4   (b) 275 ÷ 11

21(a)_____

(b)_____

22. Divide the following decimals.

   (a) 48.24 ÷ 8   (b) 15.414 ÷ 0.42

22(a)_____

(b)_____

23. Divide the following fractions.
    (Reduce answers to lowest terms.)

   (a) $\frac{1}{3} \div \frac{3}{5}$ (b) $\frac{12}{25} \div \frac{4}{5}$

23(a)_____

(b)_____

24. Divide the following mixed numbers.
    (Reduce answers to lowest terms.)

   (a) $\frac{3}{8} \div 4\frac{1}{2}$ (b) $3\frac{1}{3} \div 8$

24(a)_____

(b)_____

25. Express the following product using exponents.

   (a) x · y · x · y
   (b) x · x · y · y · y

25(a)_____

(b)_____

26. Evaluate.

   (a) $2^3 + 2$ (b) $2 \cdot 2^2 \cdot 3^2$

26(a)_____

(b)_____

27. Evaluate using the proper order of operations.

   (a) 5 x 4 - 4 x 3   (b) 3[5 - (3+1)]

27(a)_____

(b)_____

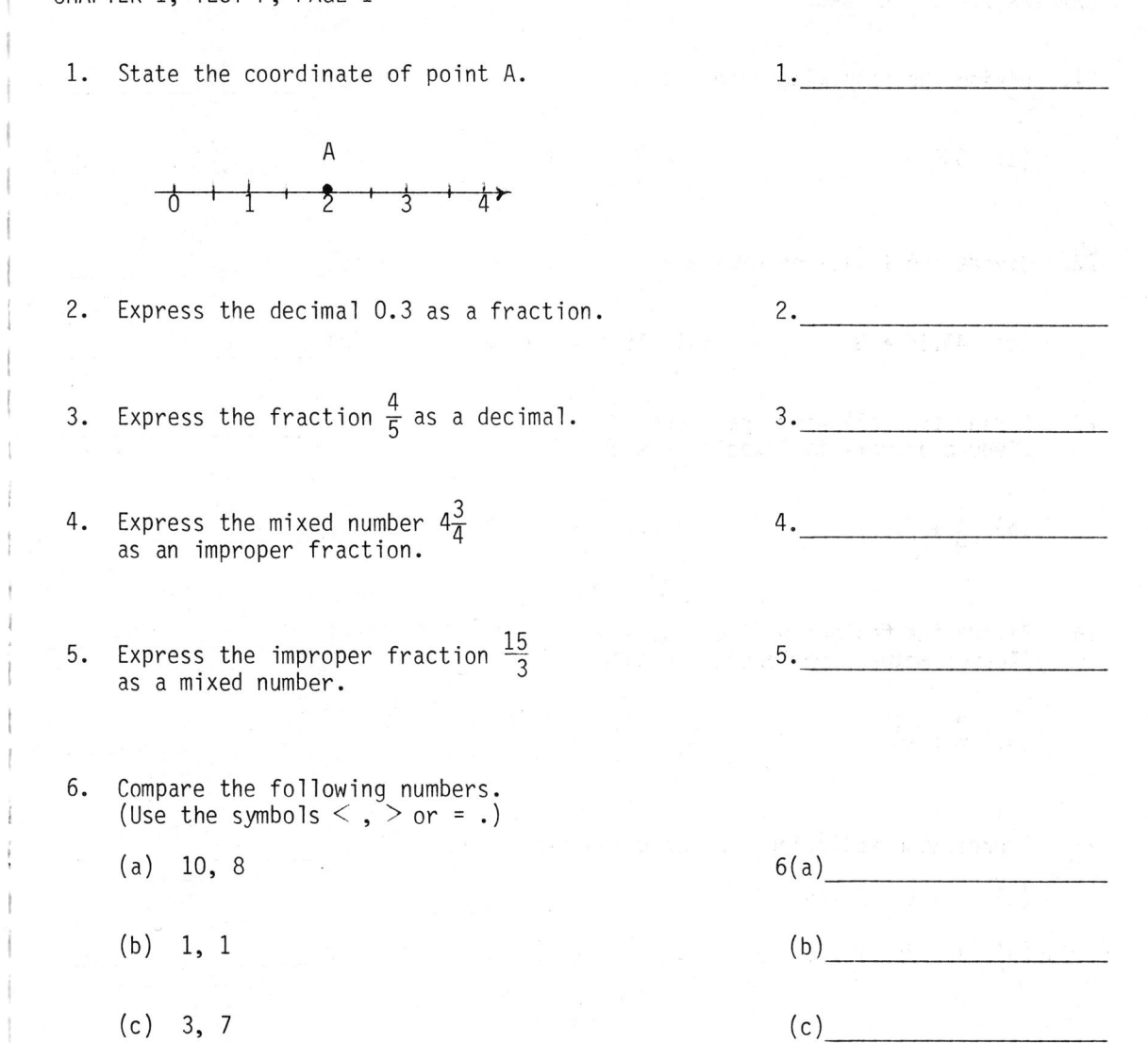

INTRODUCTORY ALGEBRA
CHAPTER 1, TEST F, PAGE 1

1. State the coordinate of point A.　　　　　　　　1. _____

   ```
        A
   +--+--•--+--+→
   0  1  2  3  4
   ```

2. Express the decimal 0.3 as a fraction.　　　　　2. _____

3. Express the fraction $\frac{4}{5}$ as a decimal.　　　　　3. _____

4. Express the mixed number $4\frac{3}{4}$　　　　　　　　4. _____
   as an improper fraction.

5. Express the improper fraction $\frac{15}{3}$　　　　　　5. _____
   as a mixed number.

6. Compare the following numbers.
   (Use the symbols <, > or = .)

   (a)　10, 8　　　　　　　　　　　　　　　　　　6(a) _____

   (b)　1, 1　　　　　　　　　　　　　　　　　　(b) _____

   (c)　3, 7　　　　　　　　　　　　　　　　　　(c) _____

7. Compare the following decimals.
   (Use the symbols <, > or = .)

   (a)　0.16, 0.167　　　　　　　　　　　　　　7(a) _____

   (b)　0.25, 0.25　　　　　　　　　　　　　　(b) _____

   (c)　3.5, 3.14　　　　　　　　　　　　　　　(c) _____

INTRODUCTORY ALGEBRA
CHAPTER 1, TEST F, PAGE 2

8. Compare the following fractions.
   (Use the symbols <, > or = .)

   (a) $\frac{9}{10}, 1\frac{1}{2}$ 8(a) _____

   (b) $\frac{2}{8}, \frac{1}{4}$ (b) _____

   (c) $1\frac{1}{9}, \frac{9}{8}$ (c) _____

9. Add the following numbers.   9(a) _____

   (a) 202 + 60   (b) 1078 + 9080   (b) _____

10. Add the following decimals.   10(a) _____

    (a) 0.125 + 0.36   (b) 0.09 + 0.99   (b) _____

11. Add the following fractions.   11(a) _____
    (Reduce answers to lowest terms.)

    (a) $\frac{1}{3} + \frac{5}{9}$ (b) $\frac{1}{9} + \frac{5}{9}$ (b) _____

12. Add the following mixed numbers.   12(a) _____
    (Reduce answers to lowest terms.)

    (a) $9\frac{5}{6} + 2\frac{1}{6}$ (b) $2\frac{2}{3} + 8\frac{2}{5}$ (b) _____

13. Subtract the following numbers.   13(a) _____

    (a) 1999 - 606   (b) 834 - 642   (b) _____

INTRODUCTORY ALGEBRA
CHAPTER 1, TEST F, PAGE 3

14. Subtract the following decimals.

    (a) 125.07 - 26.9   (b) 14.2 - 11.02

14(a)_____

(b)_____

15. Subtract the following fractions. (Reduce answers to lowest terms.)

    (a) $\frac{7}{9} - \frac{5}{9}$ (b) $\frac{3}{4} - \frac{2}{3}$

15(a)_____

(b)_____

16. Subtract the following mixed numbers. (Reduce answers to lowest terms.)

    (a) $6\frac{7}{12} - 2\frac{5}{12}$ (b) $6\frac{1}{8} - \frac{9}{16}$

16(a)_____

(b)_____

17. Multiply the following numbers.

    (a) 291 x 5   (b) 256 x 70

17(a)_____

(b)_____

18. Multiply the following decimals.

    (a) 6 x 0.02   (b) 2.4 x 3.25

18(a)_____

(b)_____

19. Multiply the following fractions. (Reduce answers to lowest terms.)

    (a) $\frac{3}{4} \times \frac{1}{2}$ (b) $\frac{3}{8} \times \frac{4}{9}$

19(a)_____

(b)_____

20. Multiply the following mixed numbers. (Reduce answers to lowest terms.)

    (a) $3\frac{1}{3} \times 3\frac{3}{5}$ (b) $2\frac{1}{6} \times 18$

20(a)_____

(b)_____

INTRODUCTORY ALGEBRA
CHAPTER 1, TEST F, PAGE 4

21. Divide the following numbers.

    (a) 7832 ÷ 8   (b) 92 ÷ 23

21(a)_____

(b)_____

22. Divide the following decimals.

    (a) 170.1 ÷ 9   (b) 17.255 ÷ 2.03

22(a)_____

(b)_____

23. Divide the following fractions.
(Reduce answers to lowest terms.)

    (a) $\frac{3}{4} \div \frac{3}{4}$ (b) $\frac{9}{10} \div \frac{6}{12}$

23(a)_____

(b)_____

24. Divide the following mixed numbers.
(Reduce answers to lowest terms.)

    (a) $3\frac{2}{5} \div 4\frac{2}{25}$ (b) $1\frac{3}{8} \div 2$

24(a)_____

(b)_____

25. Express the following product using exponents.

    (a) x · y · y · y

    (b) 3 · 3 · z · x · x · x

25(a)_____

(b)_____

26. Evaluate.

    (a) $3^2 - (2^3)$ (b) $5^2 \cdot 2^2$

26(a)_____

(b)_____

27. Evaluate using the proper order of operations.

    (a) 15 - 10 ÷ 5 + 1   (b) 2[(7-5) + 3]

27(a)_____

(b)_____

INTRODUCTORY ALGEBRA
CHAPTER 2, TEST B, PAGE 1

1. Find the value of the expression.

    (a) $6 + |-8|$

    (b) $4|-8| - |+6|$

    1(a) _____

    (b) _____

2. Identify which addition statement is illustrated by the diagram.

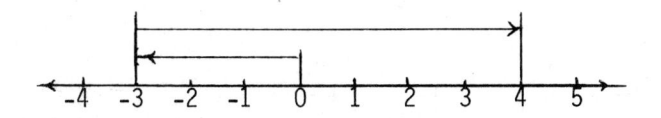

    (a) $-3 + 4$ (b) $-3 + 7$

    (c) $4 + (-3)$ (d) $3 + (-7)$

    2. _____

3. Find the sum.

    (a) $24 + (-9)$

    (b) $-9 + (-6) + 5$

    (c) $-10 + (-\frac{4}{5})$

    3(a) _____

    (b) _____

    (c) _____

4. Find the difference.

    (a) $13 - (-7)$

    (b) $-17 - (-6+2)$

    4(a) _____

    (b) _____

5. Find the product of the following numbers.

    (a) $(+7)(-4)$

    (b) $(+\frac{2}{5})(-3)$

    (c) $(-\frac{6}{11})(-\frac{5}{6})$

    (d) $(-7)[5+(-4)]$

    5(a) _____

    (b) _____

    (c) _____

    (d) _____

INTRODUCTORY ALGEBRA
CHAPTER 2, TEST B, PAGE 2

6. Evaluate the following by finding the product.

   (a) $(-5)(-4)(\frac{1}{2})$ 6(a) _____

   (b) $-5(-4)^2$ (b) _____

7. Find the quotient for the following division problems.

   (a) $-25 \div -5$ 7(a) _____

   (b) $-8 \div (-\frac{1}{2})$ (b) _____

   (c) $\frac{-80}{5}$ (c) _____

8. Perform the indicated operations.

   (a) $3(-4) + (-4)(3)$ 8(a) _____

   (b) $3 + 24 \div (-2)^3 - 2^2$ (b) _____

   (c) $\frac{3 - 5^2}{6}$ (c) _____

9. State the property that makes each statement true.

   (a) $2 + (6+8) = (2+6) + 8$ 9(a) _____

   (b) $3 + 0 = 3$ (b) _____

   (c) $3(x) + 3(y) = 3(y) + 3(x)$ (c) _____

INTRODUCTORY ALGEBRA
CHAPTER 2, TEST C, PAGE 1

1. Find the value of the expression.

    (a) $-(-7)+|-5|$ 1(a)_____

    (b) $4|-5|-|-7|$ (b)_____

2. Identify which addition statement is illustrated by the diagram.

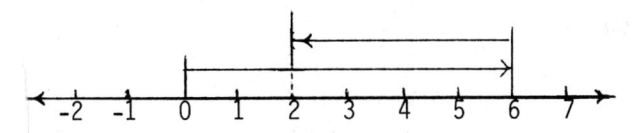

    (a) 6 + (-4)   (b) -4 + 6   2._____

    (c) 2 + (-4)   (d) 2 + 4

Find the sum.

    (a) -8 + 12   3(a)_____

    (b) 7 + (-6) + (-5)   (b)_____

    (c) $-8 + (-\frac{2}{7})$ (c)_____

Find the difference.

    (a) -4 - 12   4(a)_____

    (b) 10 - (100-25)   (b)_____

5. Find the product of the following numbers.

    (a) (-8)(+5)   5(a)_____

    (b) $(-4)(+\frac{5}{12})$ (b)_____

    (c) $(+\frac{8}{9})(-\frac{11}{3})$ (c)_____

    (d) 3[(-8)+4]   (d)_____

INTRODUCTORY ALGEBRA
CHAPTER 2, TEST C, PAGE 2

6. Evaluate the following by finding the product.

    (a) $(\frac{3}{4})(-6)(2)$ 6(a)_____

    (b) $(-5)^3$ (b)_____

7. Find the quotient for the following division problems.

    (a) $16 \div -8$ 7(a)_____

    (b) $-12 \div (+\frac{3}{4})$ (b)_____

    (c) $\frac{16}{-4}$ (c)_____

8. Perform the indicated operations.

    (a) $-2 + 18 - (3 + 9 \div (-3))$ 8(a)_____

    (b) $-2(4 + 6 \div 3)^2$ (b)_____

    (c) $\frac{-3^2 + (-2)^2}{5 - 8}$ (c)_____

9. State the property that makes each statement true.

    (a) $5 \cdot 1 = 5$ 9(a)_____

    (b) $0 \cdot x = 0$ (b)_____

    (c) $(-2) \cdot (1) = (1)(-2)$ (c)_____

INTRODUCTORY ALGEBRA
CHAPTER 2, TEST D, PAGE 1

1. Find the value of the expression.

   (a) $-(-4)+|-9|$ 1(a)_____

   (b) $3|+6|-|-12|$ (b)_____

2. Identify which addition statement is illustrated by the diagram.

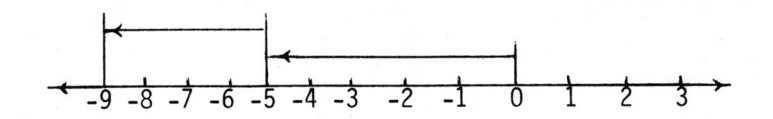

   (a) $-9 + (-5)$ (b) $- 5 + (-4)$ 2._____

   (c) $4 + (-5)$ (d) $4 + 5$

3. Find the sum.

   (a) $-15 + (-6)$ 3(a)_____

   (b) $-3 + 1 + (-16)$ (b)_____

   (c) $\frac{7}{8} + (-2)$ (c)_____

4. Find the difference.

   (a) $-6 - 9$ 4(a)_____

   (b) $(7-21) - (-18+12)$ (b)_____

5. Find the product of the following numbers.

   (a) $(-7)(-8)$ 5(a)_____

   (b) $(-5)(-\frac{1}{4})$ (b)_____

   (c) $(-\frac{2}{5})(-\frac{10}{12})$ (c)_____

   (d) $[8 (-2)+(-5)]$ (d)_____

INTRODUCTORY ALGEBRA
CHAPTER 2, TEST D, PAGE 2

6. Evaluate the following by finding the product.

    (a) $(7)(-3)(\frac{3}{7})$ 6(a) _____

    (b) $(-3)^2 (-2)^3$ (b) _____

7. Find the quotient for the following division problems.

    (a) $15 \div -3$ 7(a) _____

    (b) $-\frac{6}{10} \div (-\frac{1}{5})$ (b) _____

    (c) $\frac{-10}{-2}$ (c) _____

8. Perform the indicated operations.

    (a) $5 - 5 - 5 + 5 \div 5$ 8(a) _____

    (b) $2 + (-3) - 2(6-8)^2$ (b) _____

    (c) $\frac{4 + 2^2}{19 - 3^2}$ (c) _____

9. State the property that makes each statement true.

    (a) $-5 + 5 = 0$ 9(a) _____

    (b) $2(4-8) = 2(4) - 2(8)$ (b) _____

    (c) $(0)(-8) = (-8)(0)$ (c) _____

INTRODUCTORY ALGEBRA
CHAPTER 2, TEST E, PAGE 1

1. Find the value of the expression.

   (a) $3 + |-8|$

   (b) $7|-3| - |6+(-2)|$

1(a)_____

(b)_____

2. Identify which addition statement is illustrated by the diagram.

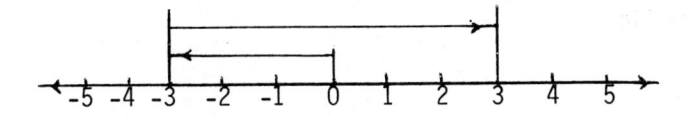

   (a) $3 + (-3)$ (b) $-6 + 3$

   (c) $-3 + (-3)$ (d) $-3 + 6$

2._____

3. Find the sum.

   (a) $-7 + (-12)$

   (b) $14 + (-8) + 2$

   (c) $-\frac{1}{2} + (-5)$

3(a)_____

(b)_____

(c)_____

4. Find the difference.

   (a) $-15 - (-8)$

   (b) $(8-4) - (6-16)$

4(a)_____

(b)_____

5. Find the product of the following numbers.

   (a) $(-12)(-3)$

   (b) $(-\frac{5}{8})(-10)$

   (c) $(-\frac{1}{2})(+\frac{4}{7})$

   (d) $5[(-12)+7]$

5(a)_____

(b)_____

(c)_____

(d)_____

INTRODUCTORY ALGEBRA
CHAPTER 2, TEST E, PAGE 2

6. Evaluate the following by finding the product.

   (a) $(10)(-\frac{1}{5})(-\frac{1}{5})$ 6(a) _____

   (b) $(-1)^4 (1)^3$ (b) _____

7. Find the quotient for the following division problems.

   (a) $-36 \div -9$ 7(a) _____

   (b) $\frac{3}{4} \div (-\frac{1}{2})$ (b) _____

   (c) $\frac{-27}{3}$ (c) _____

8. Perform the indicated operations.

   (a) $4(-3 + 5 - 6 \div 2 + 4)$ 8(a) _____

   (b) $-2(3)^2 + 4(5-6)^2$ (b) _____

   (c) $\frac{5 + (-7)^2}{-2} + 8$ (c) _____

9. State the property that makes each statement true.

   (a) $(5)(3) = (3)(5)$ 9(a) _____

   (b) $(6+2)4 = 4(2) + 4(6)$ (b) _____

   (c) $2[(-2)(4)] = [(2)(-2)(4)]$ (c) _____

INTRODUCTORY ALGEBRA
CHAPTER 2, TEST F, PAGE 1

1. Find the value of the expression.
   (a) $-(-5)+|-6|$
   (b) $5|-7|-|8+(-13)|$

   1(a)_____
   (b)_____

2. Identify which addition statement is illustrated by the diagram.

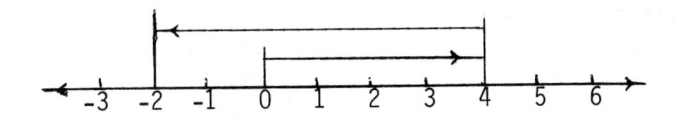

   (a) $-2 + 4$ (b) $-4 +(-6)$
   (c) $4 +(-6)$ (d) $-4 + 6$

   2._____

3. Find the sum.
   (a) $-21 + (-11)$

   (b) $-6 + (-8) + (-11)$

   (c) $-\frac{5}{9} + 7$

   3(a)_____
   (b)_____
   (c)_____

4. Find the difference.
   (a) $-5 - (-16)$
   (b) $9 - (7-10)$

   4(a)_____
   (b)_____

5. Find the product of the following numbers.
   (a) $(-5)(-10)$

   (b) $(-\frac{1}{9})(+2)$

   (c) $(-\frac{7}{9})(+\frac{3}{7})$

   (d) $(-6)[4+6]$

   5(a)_____
   (b)_____
   (c)_____
   (d)_____

INTRODUCTORY ALGEBRA
CHAPTER 2, TEST F, PAGE 2

6. Evaluate the following by finding the product.

   (a) $(-\frac{5}{6})(-\frac{3}{5})(4)$

   6(a) _____

   (b) $(-4)^2 (-1)^3$

   (b) _____

7. Find the quotient for the following division problems.

   (a) $-28 \div 4$

   7(a) _____

   (b) $-\frac{4}{9} \div (+\frac{3}{5})$

   (b) _____

   (c) $\frac{63}{-3}$

   (c) _____

8. Perform the indicated operations.

   (a) $(2+(-9)) + 2(-3) - (-4)$

   8(a) _____

   (b) $(-2 \cdot 5)^2 + (-5) - 3^2$

   (b) _____

   (c) $\dfrac{-2(-3)^2}{4 + (-6)^2}$

   (c) _____

9. State the property that makes each statement true.

   (a) $(-2) + 2 = 2 + (-2)$

   9(a) _____

   (b) $(-2) + 2 = 0$

   (b) _____

   (c) $-2 + 0 = -2$

   (c) _____

INTRODUCTORY ALGEBRA
CHAPTER 3, TEST B, PAGE 1

1. In each expression determine the number of terms, identify the terms, and state the coeffieient for each term.

   (a) 3a + b

   1(a) _____

   (b) (x + 3y)

   (b) _____

2. Write each of the following products using exponents.

   (a) x · y · x · y

   2(a) _____

   (b) -3 · x · y · y · y

   (b) _____

3. Write each of the following terms without using exponents.

   (a) x^2y

   3(a) _____

   (b) $(-2x)^4$

   (b) _____

4. Evaluate each expression for
   a = 5, b = 4, x = -2, y = 8.

   (a) ay + x

   4(a) _____

   (b) ax^2 + bx

   (b) _____

INTRODUCTORY ALGEBRA
CHAPTER 3, TEST B, PAGE 2

5. Combine like terms.

    (a) $4x - 2x$ 5(a)_____

    (b) $6x - 8x + 4y - 11y$ (b)_____

6. Remove the grouping symbols and combine like terms.

    (a) $2x - (x + y)$ 6(a)_____

    (b) $3(a + b) + 2(a - b)$ (b)_____

INTRODUCTORY ALGEBRA
CHAPTER 3, TEST C, PAGE 1

1. In each expression determine the number of terms, identify the terms, and state the coeffieient for each term.

    (a) 6xy

    (b) $\dfrac{2(x + y)}{3}$

    1(a)_____

    (b)_____

2. Write each of the following products using exponents.

    (a) $x \cdot y \cdot y$

    (b) $2 \cdot 2 \cdot x \cdot y \cdot z \cdot x$

    2(a)_____

    (b)_____

3. Write each of the following terms without using exponents.

    (a) $2x^3$

    (b) $(-3x^2)^2$

    3(a)_____

    (b)_____

4. Evaluate each expression for
   $a = 5$, $b = 4$, $x = -2$, $y = 8$.

    (a) $\dfrac{xy}{b}$

    (b) $\dfrac{-b}{2a}$

    4(a)_____

    (b)_____

INTRODUCTORY ALGEBRA
CHAPTER 3, TEST C, PAGE 2

5. Combine like terms.

   (a) 6y - 9y + 14y   5(a)_____

   (b) x + 2y + 3z + 2x - 2y - 2z   (b)_____

6. Remove the grouping symbols and combine like terms.

   (a) 2x + (y + 3x)   6(a)_____

   (b) 2(x + 2) - 3(x - 2)   (b)_____

INTRODUCTORY ALGEBRA
CHAPTER 3, TEST D, PAGE 1

1. In each expression determine the number of terms, identify the terms, and state the coeffieient for each term.

   (a) -13xyz

   1(a)_____

   (b) $\frac{x}{2} + \frac{y}{3}$

   (b)_____

2. Write each of the following products using exponents.

   (a) $x \cdot x \cdot y$

   2(a)_____

   (b) $2 \cdot x \cdot x \cdot 3 \cdot x$

   (b)_____

3. Write each of the following terms without using exponents.

   (a) $-2y^2$

   3(a)_____

   (b) $(2xy)^3$

   (b)_____

4. Evaluate each expression for $a = 5$, $b = 4$, $x = -2$, $y = 8$.

   (a) $2x - 8a$

   4(a)_____

   (b) $(\frac{-1}{2})(ab)$

   (b)_____

INTRODUCTORY ALGEBRA
CHAPTER 3, TEST D, PAGE 2

5. Combine like terms.

   (a) $3ab - ab + 2ab$ 5(a)_____

   (b) $4a - 3ab + 16a - 8ab$ (b)_____

6. Remove the grouping symbols and combine like terms.

   (a) $4y - (2x - 4y)$ 6(a)_____

   (b) $-5(x + y) + 2(2x - y)$ (b)_____

INTRODUCTORY ALGEBRA
CHAPTER 3, TEST E, PAGE 1

1. In each expression determine the number of terms, identify the terms, and state the coeffieient for each term.

   (a) $\frac{x}{3}$ 1(a)_____

   (b) $2(x + y) - 4$ (b)_____

2. Write each of the following products using exponents.

   (a) $x \cdot y \cdot z \cdot x \cdot y \cdot z$ 2(a)_____

   (b) $3 \cdot x \cdot x \cdot 5 \cdot x \cdot y \cdot y$ (b)_____

3. Write each of the following terms without using exponents.

   (a) $3x^2y^2$ 3(a)_____

   (b) $(3x^2y^2)^2$ (b)_____

4. Evaluate each expression for
   $a = 5$, $b = 4$, $x = -2$, $y = 8$.

   (a) $-3xy$ 4(a)_____

   (b) $(x - a)(x - b)$ (b)_____

INTRODUCTORY ALGEBRA
CHAPTER 3, TEST E, PAGE 2

5. Combine like terms.

    (a) $4(x + y) - 4(x + y)$ 5(a)_____

    (b) $3x + x^2 - 2x^2 + x$ (b)_____

6. Remove the grouping symbols and combine like terms.

    (a) $-(5x + 2y) - x + 3y$ 6(a)_____

    (b) $-2(2x + 3y) - 3(x - y)$ (b)_____

INTRODUCTORY ALGEBRA
CHAPTER 3, TEST F, PAGE 1

1. In each expression determine the number of terms, identify the terms, and state the coeffieient for each term.

   (a) $ax + by$ 1(a)_____

   (b) $\dfrac{x-3}{2}$ (b)_____

2. Write each of the following products using exponents.

   (a) $x \cdot x \cdot x \cdot x \cdot x \cdot y$ 2(a)_____

   (b) $(-2) \cdot x \cdot (-3) \cdot x \cdot 5 \cdot x \cdot x$ (b)_____

3. Write each of the following terms without using exponents.

   (a) $-4x^2y^3$ 3(a)_____

   (b) $(-1x^2y)^4$ (b)_____

4. Evaluate each expression for $a = 5$, $b = 4$, $x = -2$, $y = 8$.

   (a) $2a - (-b)$ 4(a)_____

   (b) $ax^2 + x - 4$ (b)_____

INTRODUCTORY ALGEBRA
CHAPTER 3, TEST F, PAGE 2

5. Combine like terms.

   (a) $6x^2 - 9x^2 + x^2$ 5(a)_____

   (b) $3x + 2x^2 - x^2 + 3x^2$ (b)_____

6. Remove the grouping symbols and combine like terms.

   (a) $x + (3x + y) - 4y$ 6(a)_____

   (b) $-2(a - 3b) + 4(3a + b)$ (b)_____

INTRODUCTORY ALGEBRA
CHAPTER 4, TEST A, PAGE 1

1. For each of the given equations, solve and check.

    (a) $3x = 9$ 1(a) _____

    (b) $\dfrac{x}{2} = 3$ (b) _____

    (c) $x + 2 = 3$ (c) _____

    (d) $x - 3 = 6$ (d) _____

2. For each of the given equations, solve and check.

    (a) $2x - 4 = 2$ 2(a) _____

    (b) $\dfrac{x}{3} + 2 = 8$ (b) _____

3. For each of the given equations, solve and check.

    (a) $2x + x = 12$ 3(a) _____

    (b) $2x - 7x = 15$ (b) _____

4. For each of the given equations, solve and check.

    (a) $4x = 7 + 2x$ 4(a) _____

    (b) $8x + 5 = 2x + 29$ (b) _____

5. For each of the given equations, solve and check.

    (a) $-(x - 5) = x + 5$ 5(a) _____

    (b) $5(2x + 3) + x = 4$ (b) _____

INTRODUCTORY ALGEBRA
CHAPTER 4, TEST A, PAGE 2

6. Solve each of the following formulas for the indicated letter.

   (a) $C = 2\pi r$, for r 6(a)_____

   (b) $P = 2\ell + 2w$, for ℓ (b)_____

7. Solve and graph the following inequalities.

   (a) $12 - 3x < 3x$ 7(a)_____

   (b) $11 + 8x \geq 2x + 35$ (b)_____

INTRODUCTORY ALGEBRA
CHAPTER 4, TEST F, PAGE 1

1. For each of the given equations, solve and check.

   (a) $6x = 3$ 1(a)_____

   (b) $\frac{x}{3} = 3$ (b)_____

   (c) $x + 7 = 10$ (c)_____

   (d) $x + 7 = 13$ (d)_____

2. For each of the given equations, solve and check.

   (a) $\frac{3}{5}x = 30$ 2(a)_____

   (b) $36 = 17 + 6x$ (b)_____

3. For each of the given equations, solve and check.

   (a) $7x - 2x = 0$ 3(a)_____

   (b) $21 = 2x - 9x$ (b)_____

4. For each of the given equations, solve and check.

   (a) $19x = 3 + 4x$ 4(a)_____

   (b) $5 - x = 15 - 17x$ (b)_____

5. For each of the given equations, solve and check.

   (a) $2(x + 1) - 2 = 16$ 5(a)_____

   (b) $3x + 3[5 + 2(x - 1)] = 0$ (b)_____

INTRODUCTORY ALGEBRA
CHAPTER 4, TEST F, PAGE 2

6. Solve each of the following formulas for the indicated letter.

    (a) $V = \ell wh$, for w 6(a) _____

    (b) $A = \frac{1}{2}h(b + c)$, for c (b) _____

7. Solve and graph the following inequalities.

    (a) $12 - 3x < 9$ 7(a) _____

    (b) $-x - 7 \geq 7x - 3$ (b) _____

INTRODUCTORY ALGEBRA
CHAPTER 5, TEST A, PAGE 1

Solve the following problems.

1. Five times a number less four, is equal to ten subtracted from twice the number. Find the number.

   1._____

2. Benny is twenty-two years older than Maria. If eight times Benny's age is equal to five times Maria's age increased by one hundred thirty-five, how old are Benny and Maria?

   2._____
   _____

3. Mary rides her bicycle from her house to her friend Jane's house at the rate of ten miles per hour. Not being the athletic type, Mary asks Jane for a ride home. If Jane drives Mary home at a rate of speed of twenty miles per hour, and the entire time Mary spends traveling is three hours, what is the distance between Mary's and Jane's houses?

   3._____

4. Albertson's Co-op sells yogurt covered almonds for $2.35 per pound and a special trail mix blend for $1.55 per pound. If a customer purchases three pounds of a mixture composed of these two ingredients for $2.03 per pound, how many pounds of each is purchased?

   4._____
   _____

5. Ralph and Elaine sold their house and moved to Florida. Once there, they invested part of the proceeds from the sale in a fledgling rock and roll pizza theatre and an established bicycle store. After one year the pizza theatre returned a 15% profit while the bicycle shop returned a 4% profit. If Ralph and Elaine invested two and one-half times as much money into the bicycle store than the pizza theatre, and the total amount of money earned in one year from these two investments was $6000, how much did they invest in each?

   5._____
   _____

INTRODUCTORY ALGEBRA
CHAPTER 5, TEST A, PAGE 2

6. Don and Nancy leave their respective residences, which are eighty-seven miles apart, and travel towards each other. Being the anxious type, Don travels at a rate of speed twelve miles per hour faster than Nancy. If it takes ninety minutes for Don and Nancy to meet, what are their respective rates of speed?

6. _____
   _____

7. JC's Bakery sells chocolate chip cookies for $2.00 per pound, peanut butter cookies for $2.26 per pound, and applesauce cookies for $1.80 per pound. For the indecisive consumer, JC's also sells a pre-packaged assortment of these cookies which consists of 0.8 pounds, 0.7 pounds, and 0.5 pounds, respectively. How much does this assortment sell for?

7. _____

8. In 1983, Larry Gilligan received $10,000 in royalties from the sale of a true romance novel which he authored. Larry invested part of this money in a savings account which earned 7% and the remaining part in a growth stock which earned 15%. If the total annual yield from these two investments was 9%, how much did he invest in each?

8. _____
   _____

INTRODUCTORY ALGEBRA
CHAPTER 5, TEST B, PAGE 1

Solve the following problems.

1. When five is added to twice a number, the sum is equal to the number less nine. Find the number.

   1._____

2. Given are two numbers, the first number twelve more than the second number. If the second number is subtracted from the first number, the difference is three times the second number. Find the numbers.

   2._____

3. Wegman's food stores sell dried sunflower seeds for $1.38 per pound and blanched cashews for $4.63 per pound. If ten pounds of a mixture of these two ingredients sell for $2.68 per pound, how many pounds of each is included in the mixture?

   3._____
   _____

4. Michael and Jane bicycle the same route each day. It takes Michael seventy-five minutes to complete the route whereas it takes Jane two hours. If Michael's rate of speed is nine miles per hour faster than Jane's, find the rate of speed for each.

   4._____
   _____

5. Bill inherited a modest sum of money. He decided to invest part of this money into a money market fund earning 12%, and the remaining part into treasury certificates earning 8%. At the end of one year, Bill received $1,200 in interest from these two investments. Find the amount Bill invested in each.

   5._____
   _____

6. A dollar change machine contains $118.60 worth of coins. If there are twenty-four more quarters than dimes and thirty-six more dimes than nickles, how many of each coin are in the machine?

   6._____
   _____
   _____

INTRODUCTORY ALGEBRA
CHAPTER 5, TEST B, PAGE 2

7. Mr. Giddis wants to invest $20,000, part at 18% and the remaining part at 10%. If he wants to receive a collective annual yield of 15%, how much must he invest at each rate?

7. _____
   _____

8. Calvin and David can run one mile in six minutes and ten minutes, respectively. One day they decide to have a race. If Calvin gives David an eighteen minute head start, how long will it take him to overtake David?

8. _____

INTRODUCTORY ALGEBRA
CHAPTER 5, TEST C, PAGE 1

Solve the following problems.

1. If four is subracted from twice a number, the result is the same as six added to twelve times the number. Find the number.

   1._____

2. A pair of pants cost $15 more than a shirt. If two pairs of pants and three shirts cost $150, find the cost of each.

   2._____
   _____

3. In a certain mini-triathlon contest, entrants are required to bicycle forty miles more than the distance of the marathon run. Robert was able to bicycle at a constant rate of speed of twenty miles per hour and run at a constant rate of speed of ten miles per hour. If he was able to complete both of these events in five hours, how long did it take him to complete each event?

   3._____
   _____

4. Claudia invested a sum of money which returned 12% at the end of one year. She also invested a second sum, $300 more than twice the first sum, which returned 15%. If the total annual income received was $612, how much money did Claudia invest at each rate?

   4._____
   _____

5. The Rochester Community Playhouse sold tickets for a stage performance. The tickets cost $2.50 if purchased in advance and $3.00 if purchased at the door. The total number of tickets sold was eight hundred fifty. If the total amount of money received from all ticket sales was $2,400, how many of each kind were sold?

   5._____
   _____

INTRODUCTORY ALGEBRA
CHAPTER 5, TEST C, PAGE 2

6. A train and a car leave Chicago, Illinois at the same time traveling in opposite directions. If the train travels at a constant rate of speed of sixty miles per hour and the car travels at a constant speed of thirty-five miles per hour, how long will it take the train and car to become eight hundred seventeen miles apart?

6. _____

7. Nan has $8,000 to invest. Looking for both a long term and short term investment, she decides to invest part of the money in a 2 1/2 year term account which has an annual yield of 15%, and the remaining part in a 90 day certificate with an annual yield of 12.5%. If Nan hopes to realize an overall annual yield of 14%, how should she divide her investment?

7. _____
  _____

8. The Goodie-Shoppe sells chocolate cremes for $2.50 per pound, butterscotch for $1.80 per pound, and chocolate covered cherries for $3.10 per pound. Jane Alberg, a candy addict, wishes to purchase one pound of chocolate cremes, three-fourths pound of butterscotch, and two pounds of chocolate covered cherries. How much per pound should the candy man charge her for this mixture?

8. _____

INTRODUCTORY ALGEBRA
CHAPTER 5, TEST D, PAGE 1

Solve the following problems.

1. The sum of nine and three times a number is equal to twice the number subtracted from nineteen. Find the number.

   1. _____

2. Given are two numbers, the second number eight less than twice the first. If three times the first number is increased by five, the sum is equal to twice the second number. Find the numbers.

   2. _____

3. Robb's Apple Farm sells Mutsu apples for $0.89 per pound and Cortland apples for $0.59 per pound. If a ten pound mix of these apples sells for $0.77 per pound, how many pounds of each apple is included in the mix?

   3. _____
   _____

4. A sum of $50,000 is invested, part at 12% and the remaining part at 8 1/2%. If the total annual income received from both investments is $5,300, how much was invested at each rate?

   4. _____
   _____

5. Driving from her home to the airport, Sylvia maintains a constant rate of speed of forty-nine miles per hour. On the return trip home, Sylvia encounters a thunderstorm and is able to travel a constant rate of speed of thirty-five miles per hour. If the round trip takes three hours, how long did it take Sylvia to get to and from the airport?

   5. _____
   _____

6. Jane took off on her bicycle at a rate of twelve miles per hour. Along her route her bicycle got a flat tire. Having no other choice, Jane walked with her bicycle back home. If she walked at constant rate of three miles per hour, and the entire ordeal took one hour, how far from home did Jane's bicycle get a flat tire?

   6. _____

INTRODUCTORY ALGEBRA
CHAPTER 5, TEST D, PAGE 2

7. A certain receipe requires beef, chicken and pork. If purchased separately, beef costs $3.10 per pound, chicken costs $2.30 per pound, and pork costs $3.30 per pound. Hoang Lee's Meat Market, however, sells a special mix of these ingredients for $3.00 per pound. If this the case, how much pork should be added to twenty pounds of beef and thirty-five pounds of chicken to maintain the selling price of $3.00 per pound?

7. _____

8. Melanie has $1000 to invest. She wishes to invest part of this amount in a savings account with a yield of 6 1/2%, and the remaining part in a trust fund earning 9%. How much should she invest in each so that her total annual yield is 8% ?

8. _____
   _____

INTRODUCTORY ALGEBRA
CHAPTER 5, TEST E, PAGE 1

Solve the following problems.

1. If three times a number is subtracted from one hundred, the result is equal to twice the number less twenty-five. What is the number?

    1._____

2. Ron is six years older than George. If three times George's age is subtracted from four times Ron's age, the result is equal to twice George's age increased by six. Find the ages of both men.

    2._____
    _____

3. Gladys invested a sum of money at 12% and a second sum, four more than twice the first sum, at 15%. If the total annual interest earned was $244.20, how much did she invest at each rate?

    3._____
    _____

4. Roses sell for $30 per dozen and carnations sell for $18 per dozen. How much would a special mix of seventy-five dozen roses and one hundred twenty-five dozen carnations cost per dozen?

    4._____

5. Two vehicles, five hundred sixty miles apart, start traveling towards each other at the same time. If the first vehicle is traveling thirty miles per hour faster than the second vehicle, how fast are the two vehicles traveling?

    5._____
    _____

6. Dried apricots sell for $2.10 per pound, dried banana chips sell for $1.85 per pound, and dried apples sell for $1.50 per pound. Five pounds of the dried apricots, two pounds of the dried banana chips, and three pounds of the dried apples are combined into one special dried fruit mixture. How much should this mixture sell for?

    6._____

INTRODUCTORY ALGEBRA
CHAPTER 5, TEST E, PAGE 2

7. Jacal, Incorporated, is considering investing $3000 into two types of investments. The first investment will yield an annual rate of 13%, whereas the second investment will return an annual rate of only 5%. How much should Jacal, Inc., invest at each rate if it wishes to earn a total annual yield of 8% ?

7._____
   _____

8. Tony can bicycle to work at a constant rate of eighteen miles per hour. One day Tony's wife, Joyce, decided to follow Tony by car. Joyce leaves two hours after Tony sets out and travels at a constant rate of speed of forty-five miles per hour. If she and Tony arrive at the same time, how many miles do they each travel?

8._____

INTRODUCTORY ALGEBRA
CHAPTER 5, TEST F, PAGE 1

Solve the following problems.

1. Four times a number increased by three is equal to five less than twice the number. Find the number.

    1._____

2. A pair of shoes is equal to five dollars less than twelve times a pair of socks. Four pairs of socks and two pairs of shoes cost sixty dollars. Find the price of each.

    2._____
    _____

3. Apple Plus cranberry juice contains 80% pure cranberry juice, whereas Ocean Mist cranberry Juice contains 45% pure cranberry juice. How many quarts of each juice must be added to a one hundred quart mixture which contains 59% pure cranberry juice?

    3._____
    _____

4. Howard Prince invests two sums of money. The first sum returned an annual yield of 5%, and the second sum, $300 more than the first sum, returned an annual yield of 8%. If the total amount of interest Mr. Prince earned from these two investments was $35.96, how much was invested at each rate?

    4._____
    _____

5. Kelly and Kristen leave their house at the same time, riding their bicycles along the same route. If Kristen bicycles at a rate of speed of eighteen miles per hour and Kelly bicycles at a rate of twenty-two miles per hour, how long will it take them to be two miles apart?

    5._____

INTRODUCTORY ALGEBRA
CHAPTER 5, TEST F, PAGE 2

6. Tickets for a Rolling Stones concert fall into three differrent classifications: general admission - $8.50; and two types of reserved seating - $12.50 and $15.00, respectively. There were three times as many general admission tickets sold than the $12.50 reserved seating tickets, and twice as many less four $15.00 reserved seating tickets than $12.50 tickets. If the total amount of money received from ticket sales was $101,396, how many tickets of each were sold?

6. _____
   _____
   _____

7. Mike and Mary left Rochester, New York, and drove south to Clarksburg, Maryland, a distance of three hundred seventy-four miles. Part of the way, they traveled at a constant rate of speed of forty-five miles per hour. The remaining part, their rate of speed was fifty-two miles per hour. If the entire trip took eight hours, how long were they traveling at each rate?

7. _____
   _____

8. Pete has $2500 to invest. He would like to invest the money into two separate accounts, one that will earn 10% and the other that will earn 5 1/2%. How much money should he invest in each account if he wishes to earn a total annual yield of 8.74% ?

8. _____
   _____

INTRODUCTORY ALGEBRA
CHAPTER 6, TEST A, PAGE 1

1. Add the following monomials.　　　　　　　　　　1(a)_____

　　(a)　4y + 2y　　　　　(b)　2x + 2x　　　　　(b)_____

　　(c)　$15x^2$　　　　　　　　　　　　　　　　(c)_____
　　　　x^2
　　　　$-5x^2$

2. Add the following polynomials.　　　　　　　　　2(a)_____

　　(a)　5 + 3x　　　　　　　　　　　　　　　　(b)_____
　　　　2 + x
　　　　　　　　　　　　　　　　　　　　　　　(c)_____
　　(b)　$-3x^2 - x + 5 + 4x - 2x^2$

　　(c)　$(2x-6) + (3x^2-1) + (5x^2-3x-4)$

3. Subtract the following monomials.　　　　　　　3(a)_____

　　(a)　　4x　　　　　(b)　-6x - (-x)　　　　　(b)_____
　　　　-(3x)

4. Subtract the following polynomials.　　　　　　4(a)_____

　　(a)　　3x - y　　　　(b)　(4-x) - (4+x)　　　(b)_____
　　　　-(2x + 2y)
　　　　　　　　　　　　　　　　　　　　　　　(c)_____
　　(c)　$(x^2-xy+y^2) - (x^2+y^2)$

5. Evaluate the following expressions.　　　　　　5(a)_____
　　　　　　　　　　　　　　　　　　　　　　　(b)_____
　　(a)　$x^2 \cdot x^3$　　　　(b)　$(a^2)^3$
　　　　　　　　　　　　　　　　　　　　　　　(c)_____

　　(c)　$(ab)^2$　　　　　(d)　$(-2x^2y)^3$　　(d)_____

INTRODUCTORY ALGEBRA
CHAPTER 6, TEST A, PAGE 2

6. Find the product of the following.

   (a) $(4a^2)(5a^3)$ (b) $7x(4x+12)$

   (c) $(x+4)(x-9)$ (d) $(2x-5)(4x^2-20x+25)$

6(a)_____
 (b)_____
 (c)_____
 (d)_____

7. Divide the following.

   (a) $\dfrac{x^7}{x^4}$ (b) $\dfrac{xy^4}{xy}$

   (c) $\dfrac{x^2}{x^3}$ (d) $\dfrac{x^2yz}{x^3y}$

7(a)_____
 (b)_____
 (c)_____
 (d)_____

8. Divide the following.

   (a) $(21ab^2) \div (-3b)$

   (b) $(7x^2+14x) \div (7x)$

   (c) $(3x^3-14x^2+17x-6) \div (x-3)$

8(a)_____
 (b)_____
 (c)_____

INTRODUCTORY ALGEBRA
CHAPTER 6, TEST D, PAGE 1

1. Add the following monomials.

   (a) $12x^3 + (-x^3)$ (b) $(-2x) + (-5x)$

   (c) $-4x$
       $+ x$

   1(a)_____
   (b)_____
   (c)_____

2. Add the following polynomials.

   (a) $12x^2 + x^2y - xy + 3$
       $\underline{2x^2 - x^2y + 3xy - 4}$

   (b) $(3x^2 + 4) + (-x^2)$

   (c) $(11x+5y) + (-11x-5y)$

   2(a)_____
   (b)_____
   (c)_____

3. Subtract the following monomials.

   (a) $-6ab$
       $\underline{-(-9ab)}$

   (b) $-x^2 - (-4x^2)$

   3(a)_____
   (b)_____

4. Subtract the following polynomials.

   (a) $4x^2 - x + 3$
       $\underline{-(3x^2 - 2x + 3)}$

   (b) $(3x^2-x) - (x-x^2)$

   (c) $(5x^2-6y^2) - (3x^2-12xy-y^2)$

   4(a)_____
   (b)_____
   (c)_____

5. Evaluate the following expressions.

   (a) $x \cdot x \cdot x$ (b) $(x^4)^4$

   (c) $(xy^3)^3$ (d) $(-5xy^2)^2$

   5(a)_____
   (b)_____
   (c)_____
   (d)_____

INTRODUCTORY ALGEBRA
CHAPTER 6, TEST D, PAGE 2

6. Find the product of the following.

   (a) $(-9x^2)(x^3)$ (b) $10x(9x-15)$

   (c) $(2a-3b)(a+5b)$ (d) $(3x+4-x^2)(5-x)$

6(a) _____
(b) _____
(c) _____
(d) _____

7. Divide the following.

   (a) $\dfrac{x^9}{x^7}$ (b) $\dfrac{y^6}{y^6}$

   (c) $\dfrac{x^2}{x^5}$ (d) $\dfrac{x^2 y^3}{x^3 y^2}$

7(a) _____
(b) _____
(c) _____
(d) _____

8. Divide the following.

   (a) $(-48x^2 y) \div (6y^3)$

   (b) $(12x^3 y^3 - 6x^2 y^2 + 18xy) \div (-6xy)$

   (c) $(x^2 - 7x + 12) \div (x-4)$

8(a) _____
(b) _____
(c) _____

INTRODUCTORY ALGEBRA
CHAPTER 6, TEST F, PAGE 1

1. Add the following monomials.

   (a) x + x + x   (b) 3z + (-7z)

   (c) $3a^3$
   $2a^3$
   $\underline{6a^3}$

   1(a)_____
   (b)_____
   (c)_____

2. Add the following polynomials.

   (a) $3x^2y + x$
   $\underline{4x^2y + 2x + y^2}$

   (b) $8x^2 + x^2 + y^2 + 3y^2$

   (c) $(x+5y) + (6x+y)$

   2(a)_____
   (b)_____
   (c)_____

3. Subtract the following monomials.

   (a) $6x^2$
   $\underline{-(-x^2)}$

   (b) $2ab - (-2ab)$

   3(a)_____
   (b)_____

4. Subtract the following polynomials.

   (a) $-8x^2 - xy$
   $\underline{-(x^2 - xy + y)}$

   (b) $x^2 + 4 - (x^2 - x)$

   (c) $(x^2 - 9x - 23) - (3x^2 - 14)$

   4(a)_____
   (b)_____
   (c)_____

5. Evaluate the following expressions.

   (a) $y^2 \cdot y^4$ (b) $(y^5)^3$

   (c) $(x^3y)^4$ (d) $(-1x^2)^4$

   5(a)_____
   (b)_____
   (c)_____
   (d)_____

INTRODUCTORY ALGEBRA
CHAPTER 6, TEST F, PAGE 2

6. Find the product of the following.

   (a) $(-\frac{1}{2}x^2y)(-4xy)$ (b) $3x(-x^2-4x+1)$

   (c) $(2x+1)(2x+1)$ (d) $(2x-1)(x^2-4x-1)$

7. Divide the following.

   (a) $\dfrac{x^9}{x^3}$ (b) $\dfrac{xyz}{xy}$

   (c) $\dfrac{y}{y^4x^2}$ (d) $\dfrac{x^5y^2}{yz^3}$

8. Divide the following.

   (a) $(-6xy^2) \div (-9x^2)$

   (b) $(3x^2-5x) \div (15x)$

   (c) $(2x^2+11x+12) \div (x+4)$

6(a) _____
(b) _____
(c) _____
(d) _____

7(a) _____
(b) _____
(c) _____
(d) _____

8(a) _____
(b) _____
(c) _____

INTRODUCTORY ALGEBRA
CHAPTER 7, TEST F, PAGE 1

1. Factor each of the following polynomials whose terms have a common factor.

   (a) $-7x^2 + 7xy - 7y^2$ 1(a)_____

   (b) $6y^2 - 4y$ (b)_____

   (c) $x(2x+1) - (2x+1)$ (c)_____

2. Find each of the indicated products by the foil method.

   (a) $(y+9)(y-1)$ 2(a)_____

   (b) $(2x+1)(x+4)$ (b)_____

   (c) $(3x-1)(3x-1)$ (c)_____

   (d) $(x+1)^2$ (d)_____

3. Factor the following trinomials.

   (a) $x^2 + 8x + 15$ 3(a)_____

   (b) $x^2 + x - 72$ (b)_____

   (c) $x^2 - x - 6$ (c)_____

   (d) $x^2 - 12x + 35$ (d)_____

INTRODUCTORY ALGEBRA
CHAPTER 7, TEST F, PAGE 2

4. Factor the following trinomials.

   (a) $3x^2 + 5x - 12$ 4(a)_____

   (b) $3x^2 + 10x + 8$ (b)_____

   (c) $10x^2 - 9x + 2$ (c)_____

   (d) $6x^2 - x - 12$ (d)_____

5. Factor the difference of two squares.

   (a) $4x^2 - 25$ 5(a)_____

   (b) $16x^2 - 100y^2$ (b)_____

6. Factor completely the following expressions.

   (a) $2y^2 - 4y - 30$ 6(a)_____

   (b) $27x^3 + 6x^2 + 3x$ (b)_____

   (c) $x^6 - x^2$ (c)_____

7. Solve and check for the roots in the following equations.

   (a) $x^2 + 5x = 0$ 7(a)_____

   (b) $4x^2 = 100$ (b)_____

   (c) $x^2 + 5x - 6 = 0$ (c)_____

   (d) $6x^2 - 7x - 20 = 0$ (d)_____

INTRODUCTORY ALGEBRA
CHAPTER 8, TEST A, PAGE 1

1. Reduce the following fractions to lowest terms.

   (a) $\dfrac{15x^3}{5x}$

   (b) $\dfrac{3x+3y}{3x-3y}$

   (c) $\dfrac{(x-7)^2}{2x-14}$

   (d) $\dfrac{x^2+2x}{x^2+5x+6}$

   (e) $\dfrac{3-x}{x^2-9}$

   1(a) _____

   (b) _____

   (c) _____

   (d) _____

   (e) _____

2. Multiply the following fractions and express the product in reduced form.

   (a) $\dfrac{3x}{8y} \cdot \dfrac{2y^2}{6x}$

   (b) $xy \cdot \dfrac{8}{x^2 y^2}$

   (c) $\dfrac{(x^2-y^2)}{8x} \cdot \dfrac{4x^3}{(x+y)^2}$

   (d) $\dfrac{x^2-49}{y^2-25} \cdot \dfrac{y-5}{x+7}$

   2(a) _____

   (b) _____

   (c) _____

   (d) _____

3. Divide the following fractions. Reduce the quotient to lowest terms.

   (a) $\dfrac{x^3}{y^2} \div \dfrac{x^2}{y}$

   (b) $\dfrac{4a^2}{7} \div 8a$

   (c) $\dfrac{x^2-y^2}{xy} \div (x-y)$

   (d) $\dfrac{x^2+4x+3}{x^2-4x-5} \div \dfrac{x+3}{x-5}$

   3(a) _____

   (b) _____

   (c) _____

   (d) _____

INTRODUCTORY ALGEBRA
CHAPTER 8, TEST A, PAGE 2

4. Combine the following fractions and simplify.

    (a) $\dfrac{2}{x-1} + \dfrac{3}{x-1}$ (b) $\dfrac{x}{x-3} - \dfrac{3}{x-3}$

    (c) $\dfrac{3x}{7} - \dfrac{x+4}{7}$

4(a) _____

(b) _____

(c) _____

5. Combine the fractions as indicated and reduce the results to lowest terms.

    (a) $\dfrac{3}{r} + \dfrac{4}{5}$ (b) $\dfrac{2x-3}{4x} - \dfrac{1+2x}{6x}$

    (c) $\dfrac{2}{3x-3y} - \dfrac{1}{2x-2y}$ (d) $a + \dfrac{a}{3}$

5(a) _____

(b) _____

(c) _____

(d) _____

6. Simplify each of the following complex fractions.

    (a) $\dfrac{x}{1 + \dfrac{1}{x}}$ (b) $\dfrac{\dfrac{x+1}{3}}{\dfrac{3x}{4}}$

6(a) _____

(b) _____

7. Solve the following equations and check.

    (a) $\dfrac{x}{21} = \dfrac{3}{7}$ (b) $\dfrac{3}{4x} + \dfrac{1}{x} = \dfrac{7}{8}$

7(a) _____

(b) _____

8. At the laundromat, using one washing machine, it takes Mary three hours to wash five loads of wash. Using three washing machines, Mary can wash all five loads in one hour. How long will it take Mary to wash five loads of wash if she used five washing mashines?

8. _____

INTRODUCTORY ALGEBRA
CHAPTER 8, TEST C, PAGE 1

1. Reduce the following fractions to lowest terms.

   (a) $\dfrac{32xy}{-24x^3y^3}$

   (b) $\dfrac{4(x+y)}{28(x+y)}$

   (c) $\dfrac{x^2-16}{4x-16}$

   (d) $\dfrac{3x^2-5x+2}{3x^2+4x-4}$

   (e) $\dfrac{x^2-25}{5-x}$

   1(a)_____

   (b)_____

   (c)_____

   (d)_____

   (e)_____

2. Multiply the following fractions and express the product in reduced form.

   (a) $\dfrac{20xy^2}{x^2} \cdot \dfrac{x^2y}{10xy}$

   (b) $\dfrac{ab}{x-y} \cdot (x^2-y^2)$

   (c) $\dfrac{2x+10}{6x} \cdot \dfrac{3x-15}{x^2-25}$

   (d) $\dfrac{x^2-9}{x+3} \cdot \dfrac{x-3}{x^2-6x+9}$

   2(a)_____

   (b)_____

   (c)_____

   (d)_____

3. Divide the following fractions. Reduce the quotient to lowest terms.

   (a) $\dfrac{12x^2y}{20ab} \div \dfrac{3xy}{4a^2b}$

   (b) $\dfrac{6a^3b^3}{8c} \div 3a^2b^2$

   (c) $\dfrac{3x+6}{(x+1)^2} \div \dfrac{x+2}{4x-4}$

   (d) $\dfrac{16x^2-y^2}{x^2-9} \div \dfrac{4x+y}{2x-6}$

   3(a)_____

   (b)_____

   (c)_____

   (d)_____

INTRODUCTORY ALGEBRA
CHAPTER 8, TEST C, PAGE 2

4. Combine the following fractions and simplify.

    (a) $\dfrac{x}{x-1} + \dfrac{3}{x-1}$ (b) $\dfrac{5x}{2} - \dfrac{x}{2}$

    (c) $\dfrac{2x}{x^2-x-2} - \dfrac{x+2}{x^2-x-2}$

4(a) _____

(b) _____

(c) _____

5. Combine the fractions as indicated and reduce the results to lowest terms.

    (a) $\dfrac{3}{x} - \dfrac{2}{5x}$ (b) $\dfrac{x+1}{4} - \dfrac{2x-1}{2}$

    (c) $\dfrac{x}{x^2-y^2} - \dfrac{3}{x+y}$ (d) $y - \dfrac{4}{y}$

5(a) _____

(b) _____

(c) _____

(d) _____

6. Simplify each of the following complex fractions.

    (a) $\dfrac{\dfrac{x^2-16}{x}}{x-4}$ (b) $\dfrac{\dfrac{3b}{5a^2}}{\dfrac{2b}{a^2}}$

6(a) _____

(b) _____

7. Solve the following equations and check.

    (a) $\dfrac{x+4}{3} = -2$ (b) $\dfrac{5}{x} = \dfrac{7}{x-4}$

7(a) _____

(b) _____

8. Using an electric typewriter, LuAnn can type a certain report in four hours. Using a word processor, Mike can type the same report in one hour. How long would it take to type this report if the two typists worked together?

8. _____

INTRODUCTORY ALGEBRA
CHAPTER 8, TEST E, PAGE 1

1. Reduce the following fractions to lowest terms.

   (a) $\dfrac{5xy^2}{30xy}$ (b) $\dfrac{9x+9y}{12x+12z}$

   (c) $\dfrac{x^2-25}{3x+15}$ (d) $\dfrac{9x^2+6x+1}{9x^2-1}$

   (e) $\dfrac{x-2}{12-6x}$

   1(a)_____

   (b)_____

   (c)_____

   (d)_____

   (e)_____

2. Multiply the following fractions and express the product in reduced form.

   (a) $6y^2 \cdot \dfrac{4}{3y}$ (b) $\dfrac{6x^2}{5y^2} \cdot \dfrac{10xy}{6x^3}$

   (c) $\dfrac{(x-4)^2}{2x^2-32} \cdot \dfrac{4x+16}{20x}$ (d) $\dfrac{x^2-4x+3}{35x^2} \cdot \dfrac{7x^2-7x}{x^2-2x+1}$

   2(a)_____

   (b)_____

   (c)_____

   (d)_____

3. Divide the following fractions. Reduce the quotient to lowest terms.

   (a) $\dfrac{13x^3}{20y} \div \dfrac{39x^2}{5y^2}$ (b) $3a \div \dfrac{1}{5a}$

   (c) $\dfrac{x^2-9}{2x+1} \div \dfrac{x+3}{4x+2}$ (d) $\dfrac{x^2-x-6}{16x^2} \div \dfrac{x^2-4}{2x^3+6x^2}$

   3(a)_____

   (b)_____

   (c)_____

   (d)_____

INTRODUCTORY ALGEBRA
CHAPTER 8, TEST E, PAGE 2

4. Combine the following fractions and simplify.

   (a) $\dfrac{13x}{15} + \dfrac{2x}{15}$ (b) $\dfrac{5x}{9x} - \dfrac{2x}{9x}$

   (c) $\dfrac{6x-5}{x^2-1} - \dfrac{5x-6}{x^2-1}$

   4(a) _____

   (b) _____

   (c) _____

5. Combine the fractions as indicated and reduce the results to lowest terms.

   (a) $\dfrac{5}{4x} + \dfrac{1}{8y}$ (b) $\dfrac{x-2}{3x} + \dfrac{x+1}{2x}$

   (c) $\dfrac{3x}{x+2} + \dfrac{5x}{x+3}$ (d) $5 + \dfrac{3}{2y-1}$

   5(a) _____

   (b) _____

   (c) _____

   (d) _____

6. Simplify each of the following complex fractions.

   (a) $\dfrac{\dfrac{x-3}{2x^2}}{\dfrac{3x+1}{4x}}$ (b) $\dfrac{\dfrac{x}{y} - 1}{\dfrac{x}{y} + 1}$

   6(a) _____

   (b) _____

7. Solve the following equations and check.

   (a) $\dfrac{x}{2} + \dfrac{x}{4} = 6$ (b) $\dfrac{5x}{4} + \dfrac{x+3}{10} = 3$

   7(a) _____

   (b) _____

8. An inexperienced apple picker can pick fifteen bushels of apples in six hours, whereas an experienced picker requires only two hours. How long would it take to pick fifteen bushels of apples if the two worked together?

   8. _____

INTRODUCTORY ALGEBRA
CHAPTER 8, TEST F, PAGE 1

1. Reduce the following fractions to lowest terms.

   (a) $\dfrac{-21x^2y^2}{28x^3y^3}$

   (b) $\dfrac{3x-3y}{3x^2}$

   (c) $\dfrac{2x^2-2}{x+1}$

   (d) $\dfrac{x^2-2x-8}{x^2-x-6}$

   (e) $\dfrac{a^2-b^2}{3b-3a}$

   1(a) _____
   (b) _____
   (c) _____
   (d) _____
   (e) _____

2. Multiply the following fractions and express the product in reduced form.

   (a) $\dfrac{3xy^2}{11x} \cdot \dfrac{55x^2}{18xy}$

   (b) $xy^2 \cdot \dfrac{4}{xy^2}$

   (c) $\dfrac{1-4x^2}{9y^2} \cdot \dfrac{y^2}{2x+1}$

   (d) $\dfrac{x^2-5x-6}{x^2+3x} \cdot \dfrac{x+3}{6-x}$

   (c) _____
   (d) _____

3. Divide the following fractions. Reduce the quotient to lowest terms.

   (a) $\dfrac{xy^2}{x^2y} \div \dfrac{x}{y^3}$

   (b) $\dfrac{x^2}{6y} \div 5xy$

   (c) $\dfrac{x^2-1}{2x+6} \div \dfrac{xy-y}{xy+3y}$

   (d) $\dfrac{x+9}{5x+10} \div \dfrac{x^2+7x-18}{30x}$

   3(a) _____
   (b) _____
   (c) _____
   (d) _____

INTRODUCTORY ALGEBRA
CHAPTER 8, TEST F, PAGE 2

4. Combine the following fractions and simplify.

   (a) $\dfrac{2x+1}{5} + \dfrac{4-2x}{5}$

   (b) $\dfrac{3}{2(x-3)} - \dfrac{x}{2(x-3)}$

   (c) $\dfrac{2a+3b}{3x} - \dfrac{a+b}{3x}$

   4(a) _____

   (b) _____

   (c) _____

5. Combine the fractions as indicated and reduce the results to lowest terms.

   (a) $\dfrac{1}{6x} - \dfrac{5}{9y}$

   (b) $\dfrac{3-4x}{2x} + \dfrac{8x+4}{8x}$

   (c) $\dfrac{x}{x^2-25} - \dfrac{1}{2x+10}$

   (d) $x + \dfrac{2x}{5}$

   5(a) _____

   (b) _____

   (c) _____

   (d) _____

6. Simplify each of the following complex fractions.

   (a) $\dfrac{\dfrac{x}{y}+3}{\dfrac{x}{y}-3}$

   (b) $\dfrac{\dfrac{4}{x}}{3-\dfrac{1}{x}}$

   6(a) _____

   (b) _____

7. Solve the following equations and check.

   (a) $\dfrac{2x}{3} - \dfrac{5x}{6} = 4$

   (b) $\dfrac{x+2}{x-1} = \dfrac{2}{3}$

   7(a) _____

   (b) _____

8. Mrs. Ferrara can pick twenty quarts of strawberries in two hours. Her great-granddaughter, Kelly, can pick twenty quarts of strawberries in three hours. How long would it take if they worked together?

   8. _____

INTRODUCTORY ALGEBRA
CHAPTER 9, TEST A, PAGE 1

1. Perform the indicated operation.

    (a) $\sqrt{x^2}$ (b) $\sqrt{25y^4x^2}$

    1(a) _____

    (b) _____

2. Simplify the following radicals.

    (a) $\sqrt{50}$ (b) $2 \cdot \sqrt{72}$

    (c) $\sqrt{x^2y^3}$ (d) $\sqrt{18x^5y^2}$

    2(a) _____

    (b) _____

    (c) _____

    (d) _____

3. Simplify the following radicals.

    (a) $\sqrt{\dfrac{9}{16}}$ (b) $\sqrt{\dfrac{8}{25}}$

    (c) $5 \cdot \sqrt{\dfrac{80}{100}}$

    3(a) _____

    (b) _____

    (c) _____

4. Combine the following radicals.

    (a) $3\sqrt{3} + 5\sqrt{3}$

    (b) $\sqrt{5} + \sqrt{2} + 3\sqrt{2} + 2\sqrt{5}$

    (c) $\sqrt{72} - \sqrt{50}$

    4(a) _____

    (b) _____

    (c) _____

INTRODUCTORY ALGEBRA
CHAPTER 9, TEST A, PAGE 2

5. Multiply the following radicals and express the product in its simplest form.

   (a) $\sqrt{2} \cdot \sqrt{8}$

   (b) $2\sqrt{2} \cdot 5\sqrt{6}$

   (c) $(2\sqrt{3})^2$

   (d) $\sqrt{2}(\sqrt{3} + \sqrt{5})$

   5(a)_____

   (b)_____

   (c)_____

   (d)_____

6. Express each quotient in its simplest radical form.

   (a) $\dfrac{\sqrt{15}}{\sqrt{3}}$ (b) $\dfrac{1}{\sqrt{2}}$

   (c) $\dfrac{\sqrt{2}}{\sqrt{6}}$ (d) $2\sqrt{\dfrac{3}{2}}$

   6(a)_____

   (b)_____

   (c)_____

   (d)_____

7. Solve and check the following radical equations.

   (a) $\sqrt{x} = 6$ (b) $\sqrt{2x} + 3 = -1$

   7(a)_____

   (b)_____

INTRODUCTORY ALGEBRA
CHAPTER 9, TEST D, PAGE 1

1. Perform the indicated operation.

    (a) $\sqrt{16y^4}$ (b) $\sqrt{x^6y^2z^4}$

2. Simplify the following radicals.

    (a) $\sqrt{44}$ (b) $\frac{1}{3}\sqrt{45}$

    (c) $\sqrt{x^3y^4}$ (d) $2\sqrt{18x^3y^3}$

3. Simplify the following radicals.

    (a) $\sqrt{\frac{9}{49}}$ (b) $\sqrt{\frac{20}{25}}$

    (c) $8\sqrt{\frac{75}{144}}$

4. Combine the following radicals.

    (a) $4\sqrt{3} + \sqrt{3}$

    (b) $\sqrt{2} + \sqrt{2} + \sqrt{3} - \sqrt{3}$

    (c) $4\sqrt{12} + 2\sqrt{3}$

1(a)_____
 (b)_____

2(a)_____
 (b)_____
 (c)_____
 (d)_____

3(a)_____
 (b)_____
 (c)_____

4(a)_____
 (b)_____
 (c)_____

INTRODUCTORY ALGEBRA
CHAPTER 9, TEST D, PAGE 2

5. Multiply the following radicals and express the product in its simplest form.

   (a) $\sqrt{7} \cdot \sqrt{7}$

   (b) $2\sqrt{2x} \cdot \sqrt{6x^3}$

   (c) $(\sqrt{2x+1})^2$

   (d) $2\sqrt{6}(3\sqrt{2} + \sqrt{3})$

5(a)_____

(b)_____

(c)_____

(d)_____

6. Express each quotient in its simplest radical form.

   (a) $\dfrac{4\sqrt{22}}{\sqrt{11}}$

   (b) $\dfrac{\sqrt{2}}{\sqrt{7}}$

   (c) $\dfrac{\sqrt{7}}{\sqrt{14}}$

   (d) $5\sqrt{\dfrac{4}{3}}$

6(a)_____

(b)_____

(c)_____

(d)_____

7. Solve and check the following radical equations.

   (a) $\sqrt{x-2} = 2$

   (b) $\sqrt{5x-1} - 8 = -1$

7(a)_____

(b)_____

INTRODUCTORY ALGEBRA
CHAPTER 9, TEST E, PAGE 1

1. Perform the indicated operation.

   (a) $\sqrt{49z^{10}}$ (b) $\sqrt{4x^4z^4}$

   1(a)_____

   (b)_____

2. Simplify the following radicals.

   (a) $\sqrt{125}$ (b) $4 \cdot \sqrt{68}$

   (c) $\sqrt{x^7}$ (d) $3\sqrt{24x^3y^5}$

   2(a)_____

   (b)_____

   (c)_____

   (d)_____

3. Simplify the following radicals.

   (a) $6\sqrt{\dfrac{4}{9}}$ (b) $\sqrt{\dfrac{18}{36}}$

   (c) $3\sqrt{\dfrac{88}{81}}$

   3(a)_____

   (b)_____

   (c)_____

4. Combine the following radicals.

   (a) $5\sqrt{2} + 7\sqrt{2}$

   (b) $2\sqrt{5} + 2\sqrt{6} - \sqrt{5} + \sqrt{6}$

   (c) $\sqrt{32} - 3\sqrt{50}$

   4(a)_____

   (b)_____

   (c)_____

INTRODUCTORY ALGEBRA
CHAPTER 9, TEST E, PAGE 2

5. Multiply the following radicals and express the product in its simplest form.

   (a) $\sqrt{2}\ \ \sqrt{50}$

   (b) $-5\sqrt{x^2}\ \ 2\sqrt{x^3}$

   (c) $(5\sqrt{3x})^2$

   (d) $\sqrt{2}\,(\sqrt{8} - 2\sqrt{2} + 3)$

5(a)_____

(b)_____

(c)_____

(d)_____

6. Express each quotient in its simplest radical form.

   (a) $\dfrac{2\sqrt{10}}{3\sqrt{5}}$ (b) $\dfrac{3}{\sqrt{5}}$

   (c) $\dfrac{60}{3\sqrt{8}}$ (d) $10\sqrt{\dfrac{7}{20}}$

6(a)_____

(b)_____

(c)_____

(d)_____

7. Solve and check the following radical equations.

   (a) $\sqrt{\dfrac{x}{6}} = 3$ (b) $\sqrt{\dfrac{4x}{3}} - 1 = 3$

7(a)_____

(b)_____

INTRODUCTORY ALGEBRA
CHAPTER 9, TEST F, PAGE 1

1. Perform the indicated operation.

   (a) $\sqrt{64x^6}$ (b) $\sqrt{9x^2y^4z^6}$

   1(a)_____

   (b)_____

2. Simplify the following radicals.

   (a) $\sqrt{32}$ (b) $-\frac{2}{3}\sqrt{27}$

   (c) $\sqrt{x^2y^4z}$ (d) $2\sqrt{9x^2y^5}$

   2(a)_____

   (b)_____

   (c)_____

   (d)_____

3. Simplify the following radicals.

   (a) $\sqrt{\dfrac{16}{36}}$ (b) $\sqrt{\dfrac{8}{9}}$

   (c) $14\sqrt{\dfrac{72}{49}}$

   3(a)_____

   (b)_____

   (c)_____

4. Combine the following radicals.

   (a) $14\sqrt{3} - 7\sqrt{3}$

   (b) $10\sqrt{7} + 8\sqrt{7} - 5\sqrt{7} - 3\sqrt{7}$

   (c) $2\sqrt{45} - 3\sqrt{20}$

   4(a)_____

   (b)_____

   (c)_____

INTRODUCTORY ALGEBRA
CHAPTER 9, TEST F, PAGE 2

5. Multiply the following radicals and express the product in its simplest form.

   (a) $\sqrt{x} \cdot \sqrt{x}$

   (b) $\sqrt{27a} \cdot \sqrt{3a}$

   (c) $(2\sqrt{x^3})^2$

   (d) $2\sqrt{3}(3\sqrt{5} - 2\sqrt{20} - \sqrt{45})$

5(a)_____

(b)_____

(c)_____

(d)_____

6. Express each quotient in its simplest radical form.

   (a) $\dfrac{15\sqrt{10}}{5\sqrt{2}}$ (b) $\dfrac{2\sqrt{5}}{\sqrt{6}}$

   (c) $\dfrac{\sqrt{6}}{\sqrt{5}}$ (d) $4\sqrt{\dfrac{9}{24}}$

6(a)_____

(b)_____

(c)_____

(d)_____

7. Solve and check the following radical equations.

   (a) $5\sqrt{x} = -2$ (b) $\sqrt{4y-3} - 2 = 1$

7(a)_____

(b)_____

INTRODUCTORY ALGEBRA
CHAPTER 10, TEST A, PAGE 1

1. Label the following points on the given set of axes.

    A (3, 0)

    B (2, 4)

    C (-3, 0)

    D (4, -1)

    E (0, 2)

    F (-4, 1)

    G (-3, -3)

    H (0, -5)

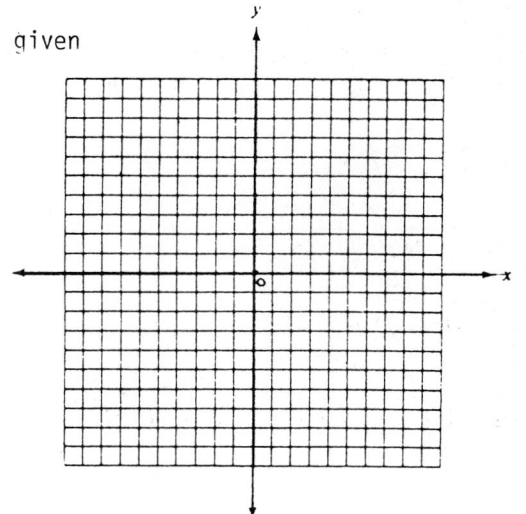

2. Name the coordinates of each of the following points.

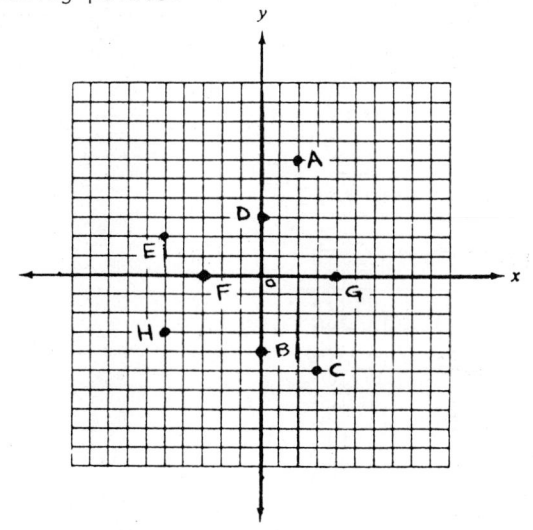

    2. A _____

    B _____

    C _____

    D _____

    E _____

    F _____

    G _____

    H _____

3. For each equation, find corresponding y-values for x = 0, 2, and -1.

    (a) y = 6 + x   (b) 2x - y = 6

    3(a) _____

    (b) _____

INTRODUCTORY ALGEBRA
CHAPTER 10, TEST A, PAGE 2

4. Graph each of the following equations on the given set of axes.

   (a) $x = y$ 4(a)

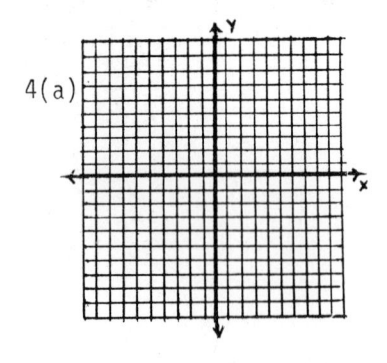

   (b) $y = 2x + 3$ 4(b)

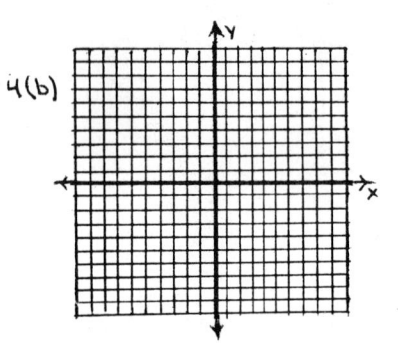

   (c) $x = -2$ 4(c)

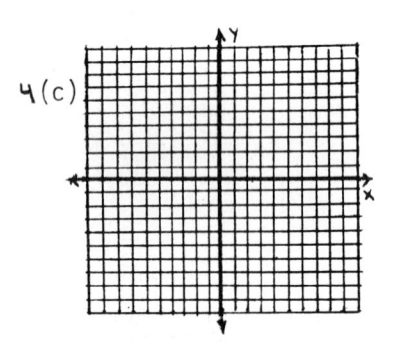

5. Graph the following lines using the intercept method.

   (a) $y + x = 4$ 5(a)

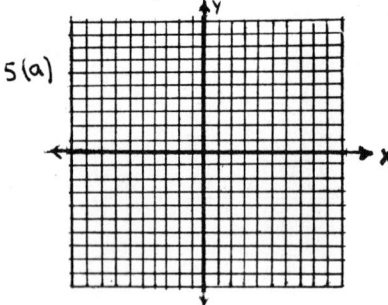

   (b) $3y = 2x + 6$ 5(b)

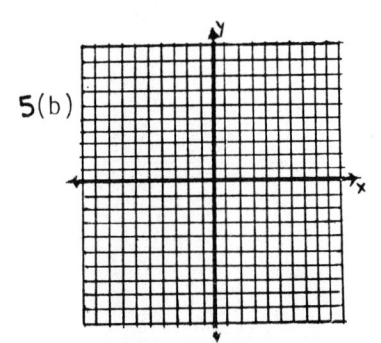

INTRODUCTORY ALGEBRA
CHAPTER 10, TEST A, PAGE 3

6. Find the slope of the line that passes through the following pairs of points.

    (a) (3, 3), (-1, -1)   6(a)_____

    (b) (1, 0), (-1, 4)   (b)_____

    (c) (3, 2), (3, -1)   (c)_____

7. Graph the line through the given point with the given slope.

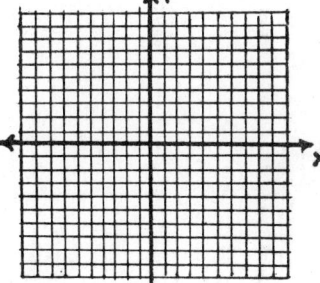

    (a) (0, 0); slope = -1   7(a)

    (b) (-1, -4); slope = 2/3   7(b)

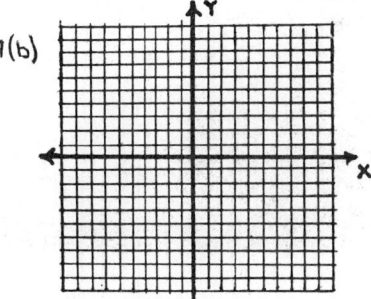

8. Find the slope and y-intercept for each of the following lines.

    (a) $y = -2x$ 8(a)_____

    (b) $3x - 4y = 7$ (b)_____

INTRODUCTORY ALGEBRA
CHAPTER 10, TEST A, PAGE 4

9. Write the equation of the line, given the following information.

   (a) The slope is -1 and passes through the point (0, -2).

   9(a) _____

   (b) The line passes through the points (0, 0) and (-2, 3).

   (b) _____

10. Graph the following inequalities.

    (a) $y > 3$

    10(a)

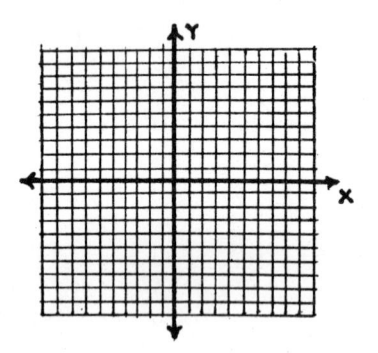

    (b) $x - 2y \leq 4$

    10(b)

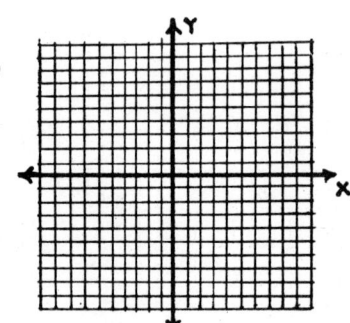

INTRODUCTORY ALGEBRA
CHAPTER 10, TEST B, PAGE 1

1. Label the following points on the given set of axes.

    A (0, 1)

    B (-4, 0)

    C (0, -1)

    D (3, 5)

    E (3, -2)

    F (4, 0)

    G (-1, -6)

    H (-1, 5)

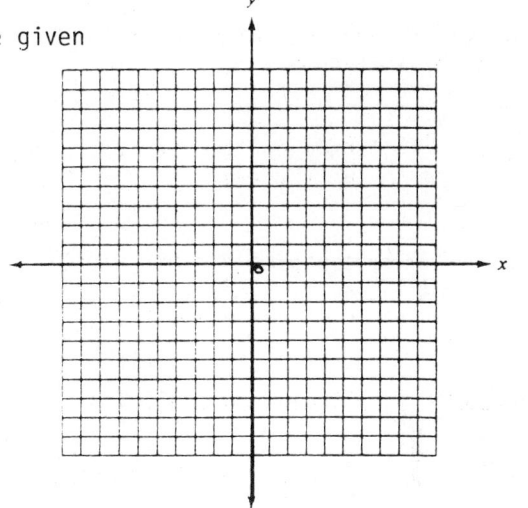

2. Name the coordinates of each of the following points.

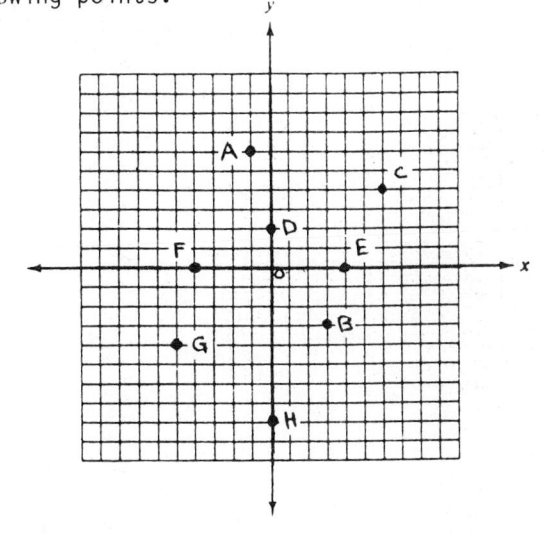

    2. A _____

       B _____

       C _____

       D _____

       E _____

       F _____

       G _____

       H _____

3. For each equation, find corresponding y-values for x = 0, 2, and -1.

    (a) $y = 4 + x$ (b) $2x + y = 7$

    3(a) _____

    (b) _____

INTRODUCTORY ALGEBRA
CHAPTER 10, TEST B, PAGE 2

4. Graph each of the following equations on the given set of axes.

   (a) $y = -x$

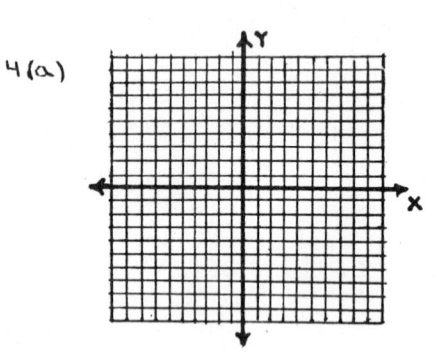

   4(a)

   (b) $y = 2x - 3$

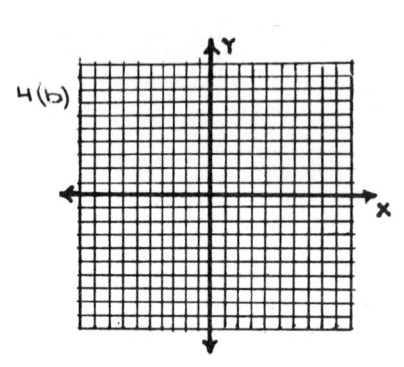

   4(b)

   (c) $y = 4$

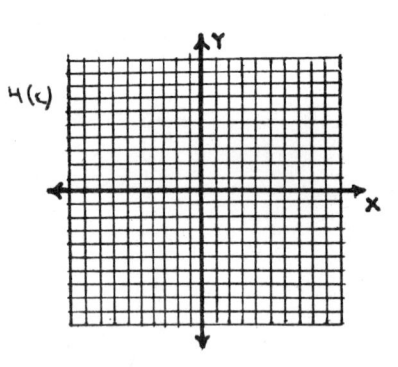

   4(c)

5. Graph the following lines using the intercept method.

   (a) $y - x = 5$

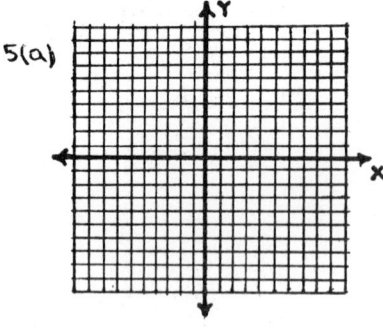

   5(a)

   (b) $4y = 3x - 12$

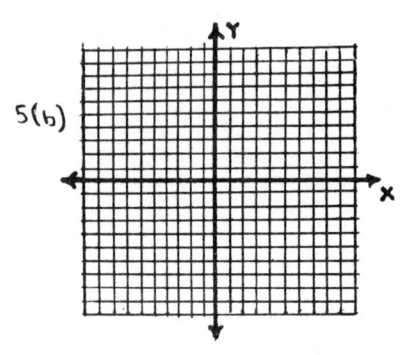

   5(b)

INTRODUCTORY ALGEBRA
CHAPTER 10, TEST B, PAGE 3

6. Find the slope of the line that passes through the following pairs of points.

   (a) (2, 5), (0, 1)   6(a)_____

   (b) (-1, -1), (5, -3)   (b)_____

   (c) (4, 2), (-1, 2)   (c)_____

7. Graph the line through the given point with the given slope.

   (a) (0, 4); slope = $-\frac{3}{2}$ 7(a)

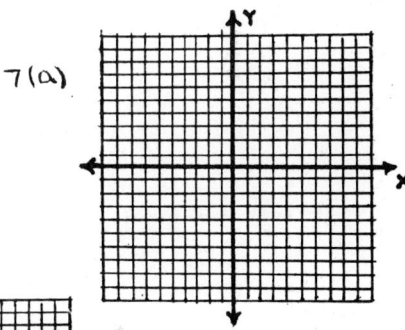

   (b) (-2, -3); slope = $\frac{3}{5}$ 7(b)

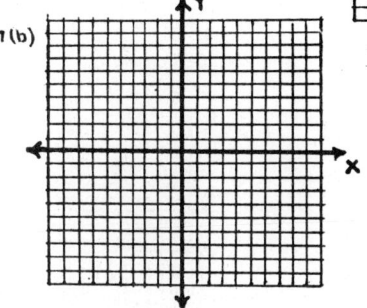

8. Find the slope and y-intercept for each of the following lines.

   (a) y = 3x - 4   8(a)_____

   (b) 3x + 2y = 6   (b)_____

INTRODUCTORY ALGEBRA
CHAPTER 10, TEST B, PAGE 4

9. Write the equation of the line, given the following information.

   (a) The slope is 1/2 and passes through the point (4, 2).

   9(a) _____

   (b) The line passes through the points (1, 3) and (3, 7).

   (b) _____

10. Graph the following inequalities.

   (a) $x + y < 1$

   10(a)

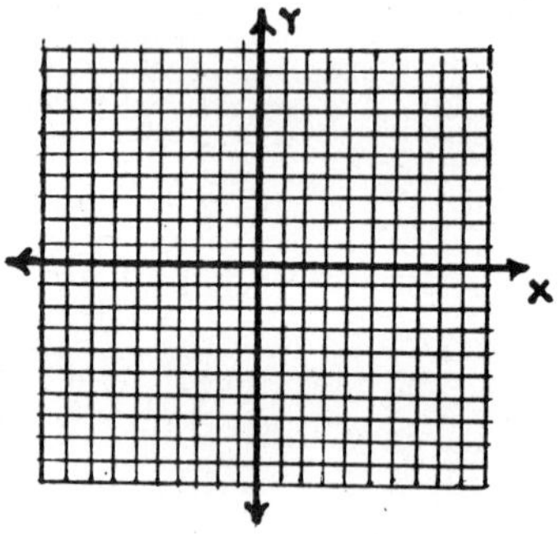

   (b) $y > 2x + 1$

   10(b)

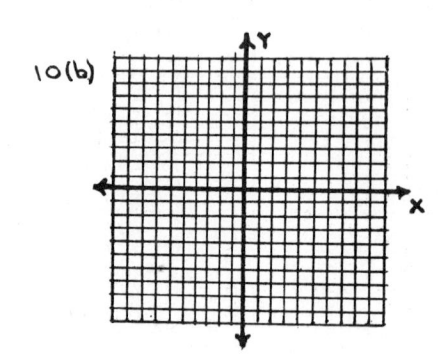

INTRODUCTORY ALGEBRA
CHAPTER 10, TEST C, PAGE 1

1. Label the following points on the given set of axes.

   A (0, -3)

   B (2, 6)

   C (1, 0)

   D (2, -5)

   E (-2, 4)

   F (0, 3)

   G (-6, 6)

   H (-4, -1)

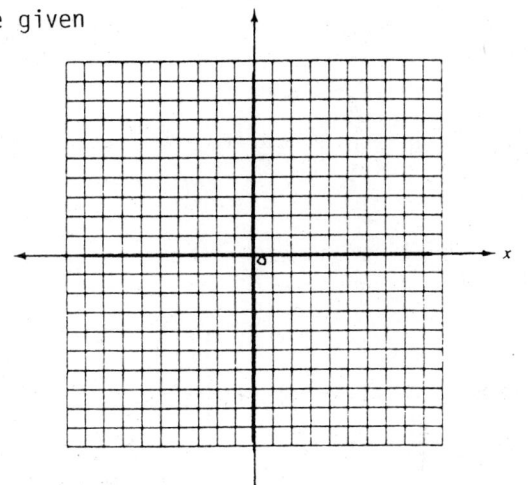

2. Name the coordinates of each of the following points.

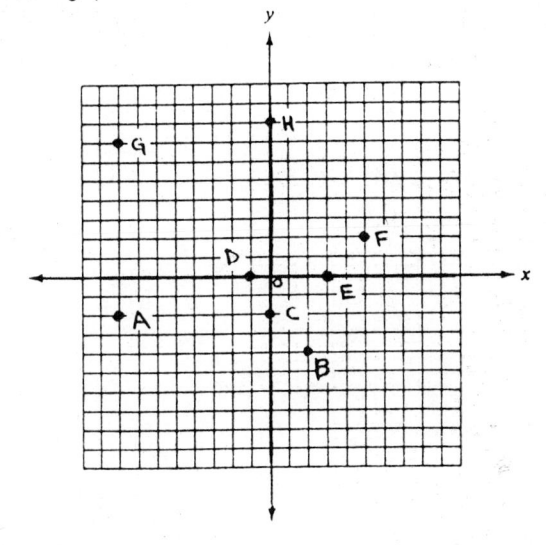

   2. A _____

   B _____

   C _____

   D _____

   E _____

   F _____

   G _____

   H _____

3. For each equation, find corresponding y-values for x = 0, 2, and -2.

   (a) y = 3x   (b) 2y - 4x = 0

   3(a) _____

   (b) _____

INTRODUCTORY ALGEBRA
CHAPTER 10, TEST C, PAGE 2

4. Graph each of the following equations on the given set of axes.

   (a) $y = x + 2$

                                             4(a)

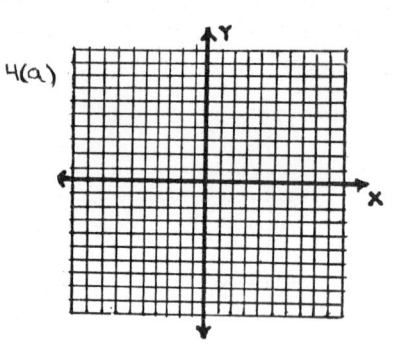

   (b) $y = 3x + 2$ 4(b)

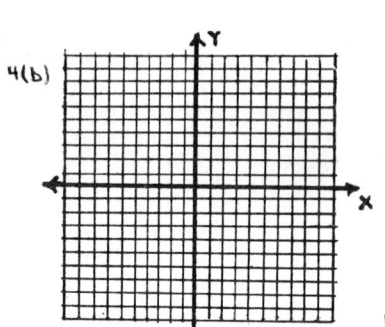

   (c) $x - 2 = 0$ 4(c)

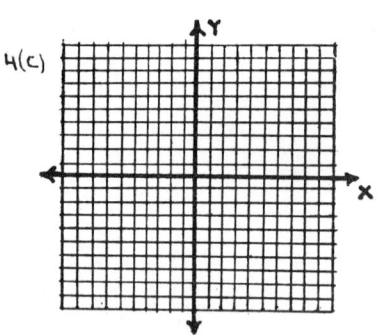

5. Graph the following lines using the intercept method.

   (a) $y + x = -3$ 5(a)

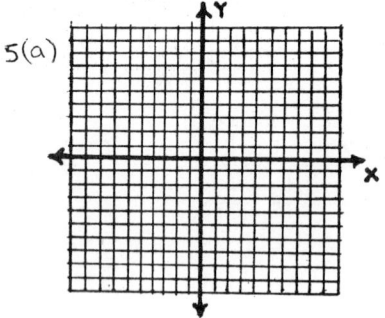

   (b) $2y - 3x = 6$ 5(b)

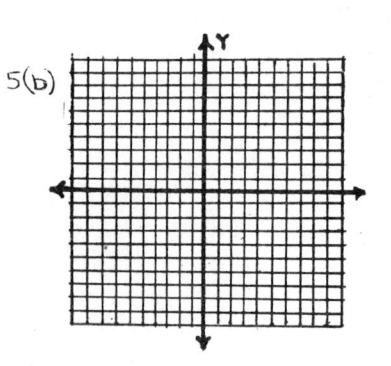

INTRODUCTORY ALGEBRA
CHAPTER 10, TEST C, PAGE 3

6. Find the slope of the line that passes through the following pairs of points.

   (a) (-3, -3), (4, -1)   6(a)_____

   (b) (1, -1), (-1, 7)   (b)_____

   (c) (-2, 5), (-2, -2)   (c)_____

7. Graph the line through the given point with the given slope.

   (a) (-2, 4); slope = $-\frac{1}{2}$

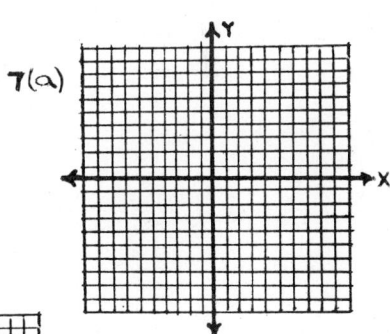

   (b) (0, 0); slope = 3

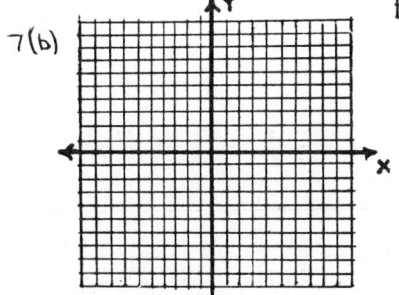

8. Find the slope and y-intercept for each of the following lines.

   (a) y = -2x + 3   8(a)_____

   (b) x + 2y = 6   (b)_____

INTRODUCTORY ALGEBRA
CHAPTER 10, TEST C, PAGE 4

9. Write the equation of the line, given the following information.

    (a) The slope is 2 and passes through the point (1, 4).

    9(a)_____

    (b) The line passes through the points (0, -1) and (6, 8).

    (b)_____

10. Graph the following inequalities.

    (a) $3x + y > -4$

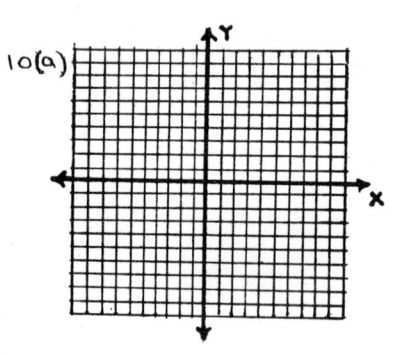

    (b) $2x + 6y < 0$

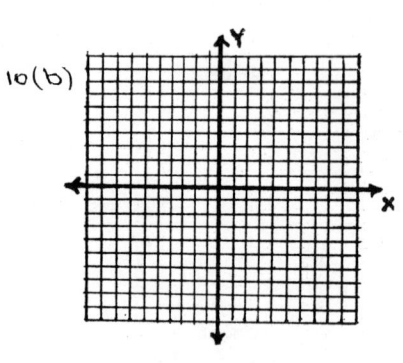

INTRODUCTORY ALGEBRA
CHAPTER 10, TEST D, PAGE 1

1. Label the following points on the given set of axes.

   A (0, 5)

   B (-1, 0)

   C (4, -1)

   D (-3, 3)

   E (5, 2)

   F (2, 0)

   G (-5, 0)

   H (0, -4)

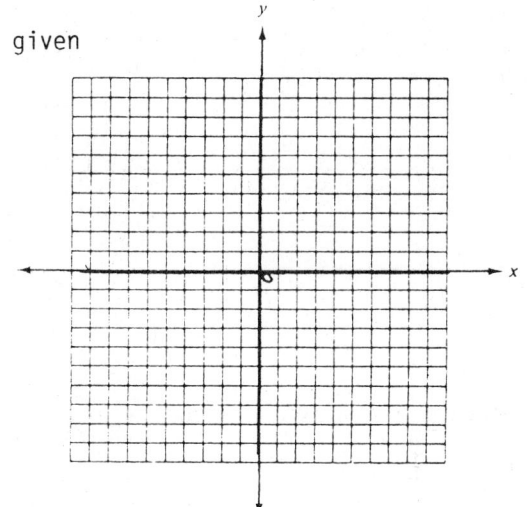

2. Name the coordinates of each of the following points.

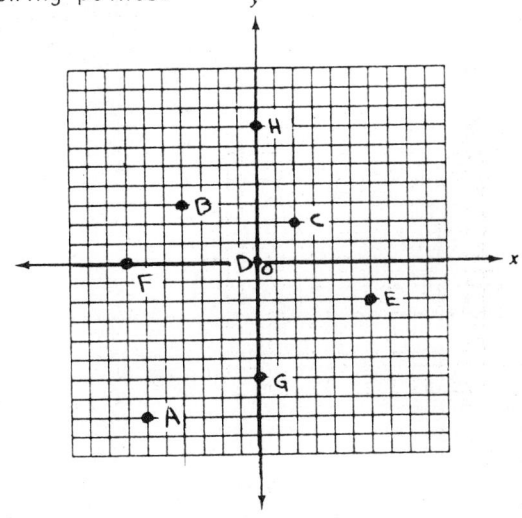

   2. A _____

   B _____

   C _____

   D _____

   E _____

   F _____

   G _____

   H _____

3. For each equation, find corresponding y-values for x = 0, 2, and -1.

   (a) $y = 2x$

   (b) $2x - y = 1$

   3(a) _____

   (b) _____

INTRODUCTORY ALGEBRA
CHAPTER 10, TEST D, PAGE 2

4. Graph each of the following equations on the given set of axes.

   (a) $x - y = 2$

   4(a)

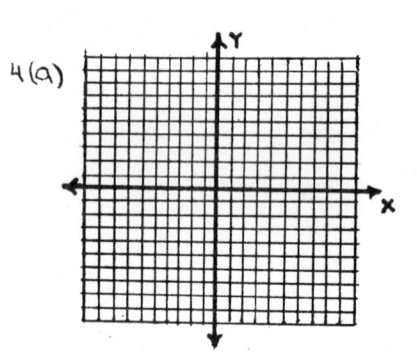

   (b) $y = 2x$

   4(b)

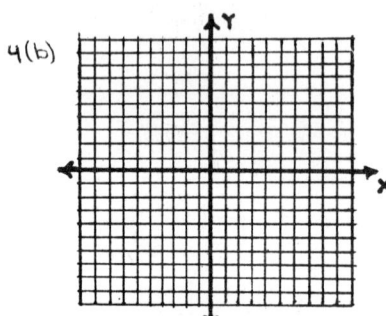

   (c) $y + 3 = 0$

   4(c)

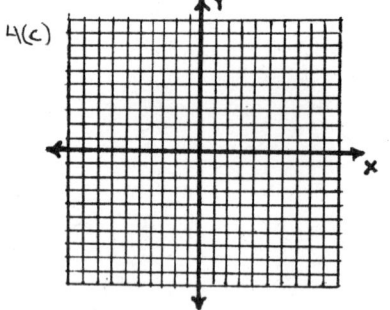

5. Graph the following lines using the intercept method.

   (a) $y - x = -4$

   5(a)

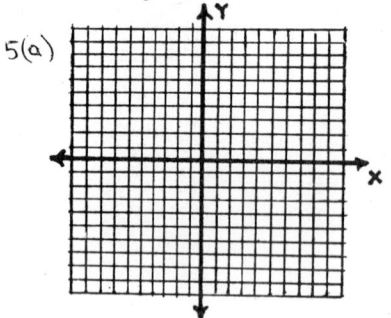

   (b) $3x + 4y = 12$

   5(b)

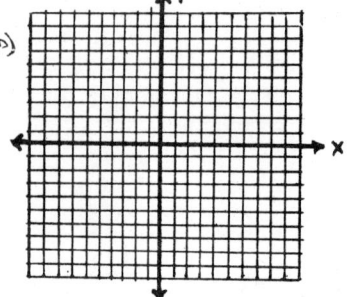

INTRODUCTORY ALGEBRA
CHAPTER 10, TEST D, PAGE 3

6. Find the slope of the line that passes through the following pairs of points.

    (a) (-3, -1), (0, 5)   6(a)_____

    (b) (-1, 5), (4, 0)   (b)_____

    (c) (0, -3), (4, -3)   (c)_____

7. Graph the line through the given point with the given slope.

    (a) (1, 4); slope = $\frac{1}{3}$

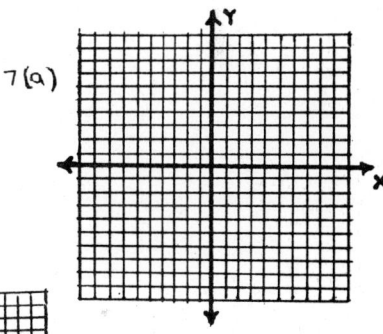

    (b) (0, -3); slope = $\frac{2}{3}$

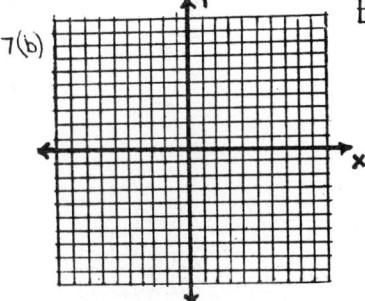

8. Find the slope and y-intercept for each of the following lines.

    (a) 2y = 3x - 6   8(a)_____

    (b) 4y - 2x = 0   (b)_____

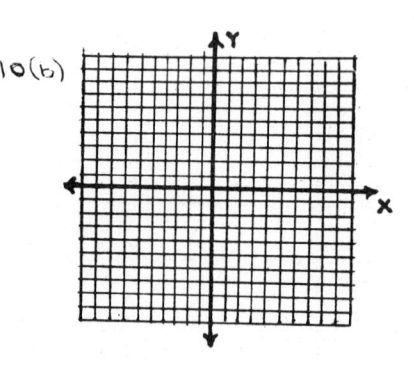

INTRODUCTORY ALGEBRA

CHAPTER 10, TEST D, PAGE 4

9. Write the equation of the line, given the following information.

   (a) The slope is 2/3 and passes through the point (-6, 4).

   9(a)_____

   (b) The line passes through the points (-1, -2) and (-2, -5).

   (b)_____

10. Graph the following inequalities.

   (a) $x - y > 2$

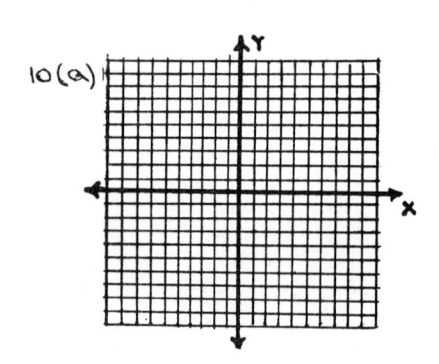

   (b) $8x - 2y > 0$

INTRODUCTORY ALGEBRA
CHAPTER 10, TEST E, PAGE 1

1. Label the following points on the given set of axes.

   A (1, -4)

   B (0, 3)

   C (3, 4)

   D (-4, -3)

   E (0, -6)

   F (-4, 1)

   G (3, 0)

   H (-4, 0)

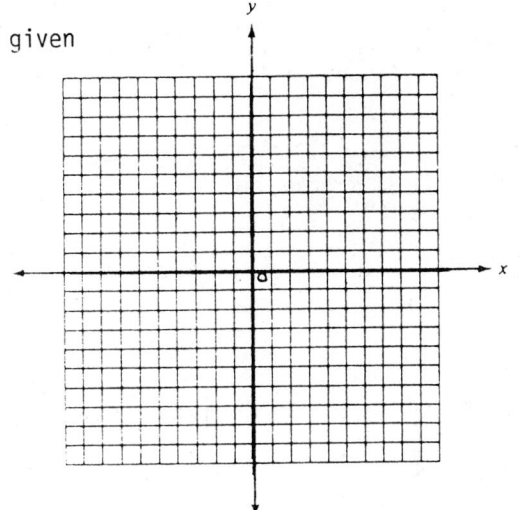

2. Name the coordinates of each of the following points.

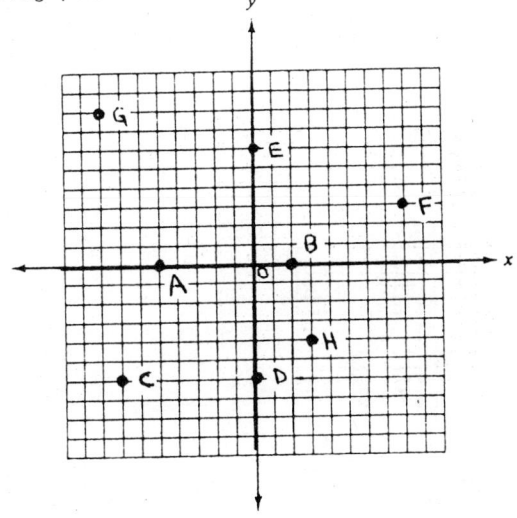

   2. A _____
   B _____
   C _____
   D _____
   E _____
   F _____
   G _____
   H _____

3. For each equation, find corresponding y-values for x = 0, 2, and -2.

   (a) y = x   (b) 3x + 2y = 6

   3(a) _____

   (b) _____

INTRODUCTORY ALGEBRA
CHAPTER 10, TEST E, PAGE 2

4. Graph each of the following equations on the given set of axes.

   (a) y = x - 3

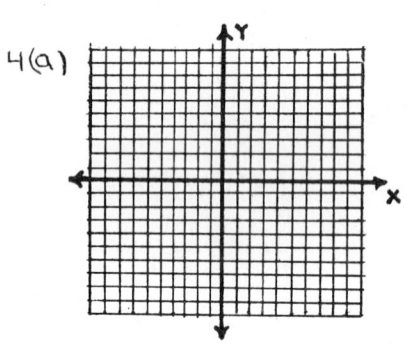

   4(a)

   (b) y = -2x + 3

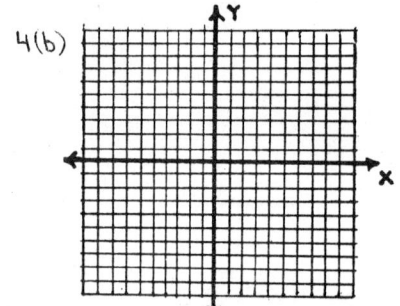

   4(b)

   (c) x + 2 = 1

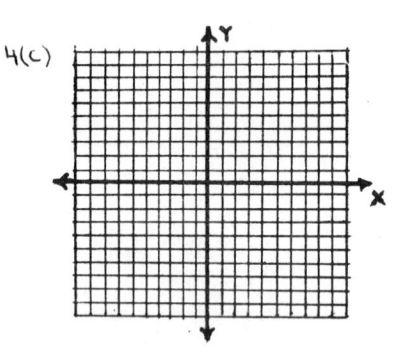

   4(c)

5. Graph the following lines using the intercept method.

   (a) y + 2x = 6

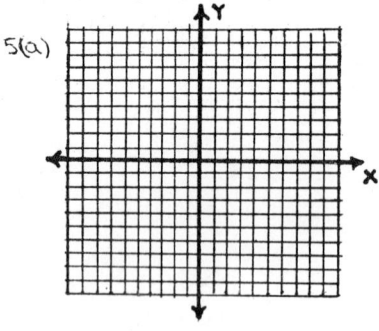

   5(a)

   (b) 5x + 3y = 15

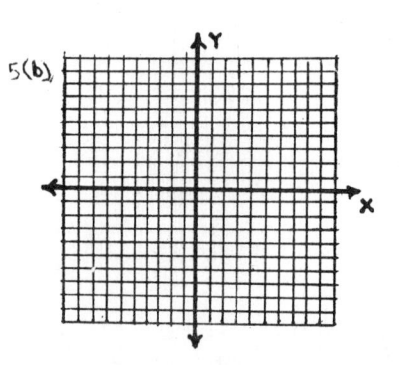

   5(b)

INTRODUCTORY ALGEBRA
CHAPTER 10, TEST E, PAGE 3

6. Find the slope of the line that passes through the following pairs of points.

   (a) (1, -3), (4, 3)   6(a)_____

   (b) (-3, 3), (2, -2)   (b)_____

   (c) (0, 5), (0, -1)   (c)_____

7. Graph the line through the given point with the given slope.

   (a) (0, 0); slope = $\frac{2}{3}$

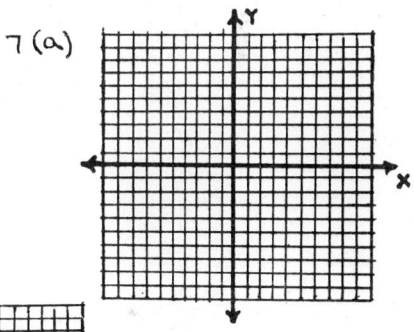

   7(a)

   (b) (-1, 4); slope = -3

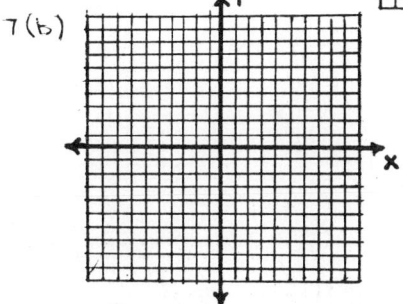

   7(b)

8. Find the slope and y-intercept for each of the following lines.

   (a) $3y = -2x + 6$ 8(a)_____

   (b) $2x + y = 7$ (b)_____

INTRODUCTORY ALGEBRA
CHAPTER 10, TEST E, PAGE 4

9. Write the equation of the line, given the following information.

    (a) The slope is 0 and passes through the point (2, -5).   9(a)_____

    (b) The line passes through the points (-1, -1) and (3, 3).   (b)_____

10. Graph the following inequalities.

    (a) x > 2   10(a)

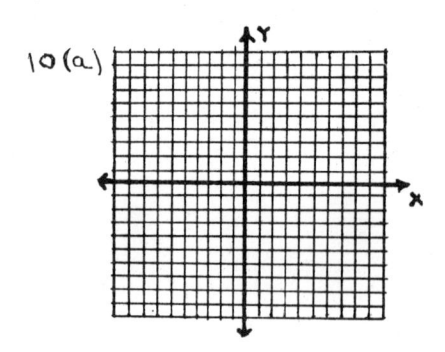

    (b) x - 3y < 4   10(b)

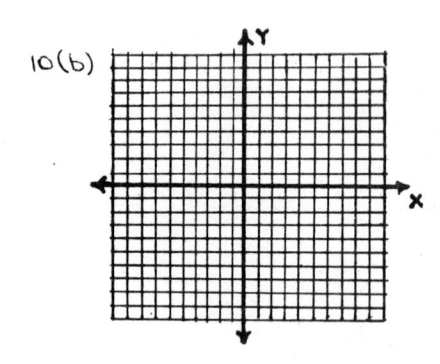

INTRODUCTORY ALGEBRA
CHAPTER 10, TEST F, PAGE 1

1. Label the following points on the given set of axes.

   A (-5, 0)

   B (0, -3)

   C (4, 3)

   D (6, -1)

   E (0, 4)

   F (-4, -3)

   G (-2, 0)

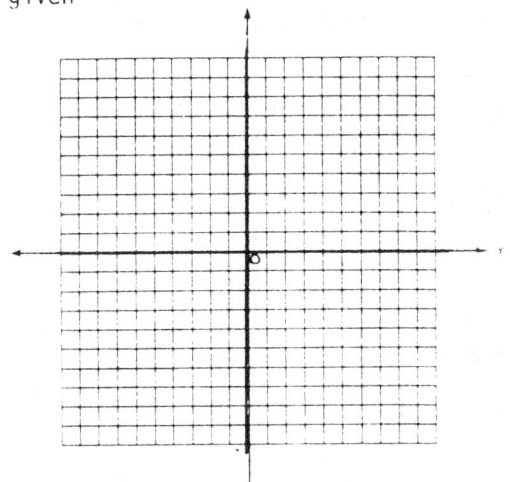

   H (5, 0)

2. Name the coordinates of each of the following points.

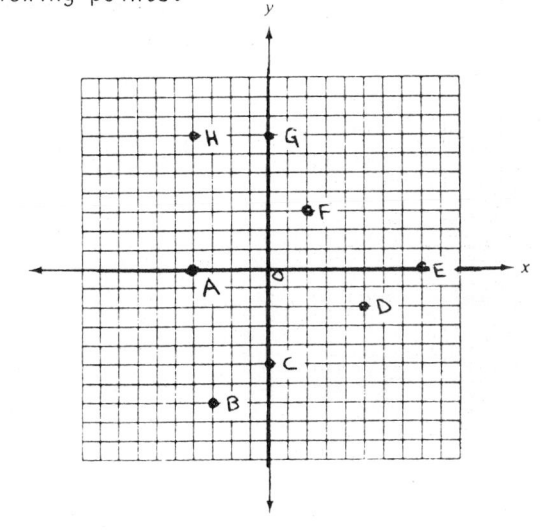

   2. A _____

   B _____

   C _____

   D _____

   E _____

   F _____

   G _____

   H _____

3. For each equation, find corresponding y-values for x = 0, 2, and -1.

   (a) $y = x - 1$ (b) $4y - 2x = 0$

   3(a) _____

   (b) _____

INTRODUCTORY ALGEBRA
CHAPTER 10, TEST F, PAGE 2

4. Graph each of the following equations on the given set of axes.

   (a) $x = y - 3$

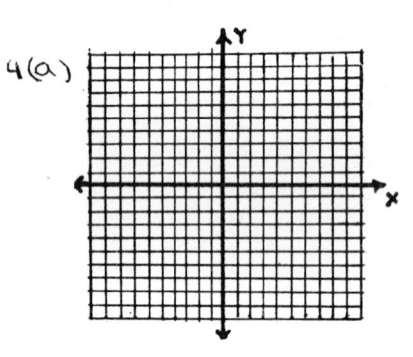

   4(a)

   (b) $y = -2x - 3$

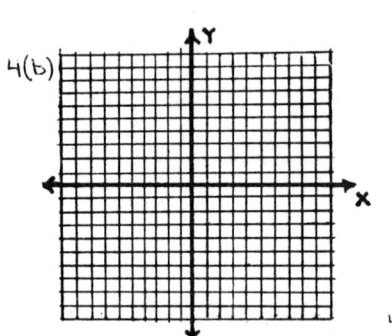

   4(b)

   (c) $y + 1 = 3$

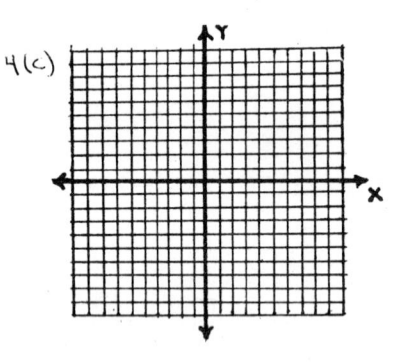

   4(c)

5. Graph the following lines using the intercept method.

   (a) $2y + x = -6$

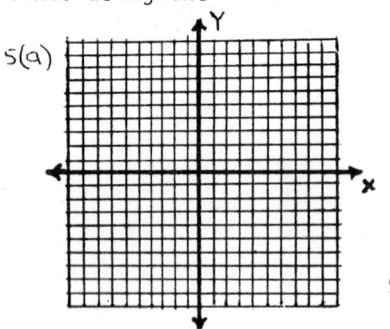

   5(a)

   (b) $3y = 5x - 15$

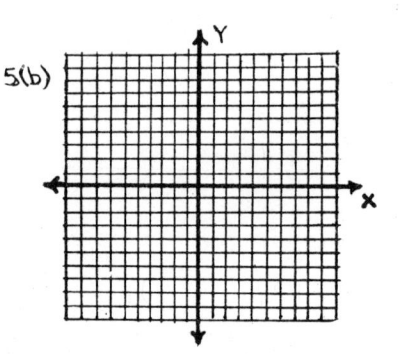

   5(b)

INTRODUCTORY ALGEBRA
CHAPTER 10, TEST F, PAGE 3

6. Find the slope of the line that passes through the following pairs of points.

   (a) (-2, 2), (4, 4)   6(a) _____

   (b) (0, 0), (-3, -6)   (b) _____

   (c) (7, 0), (-2, 0)   (c) _____

7. Graph the line through the given point with the given slope.

   (a) (1, 1); slope = $\frac{3}{4}$

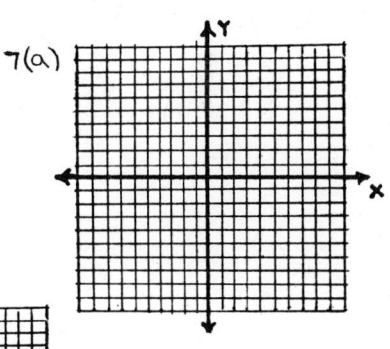

   (b) (0, 5); slope = $-\frac{2}{5}$

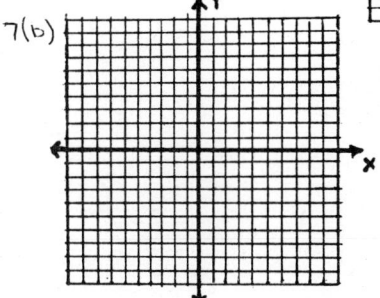

8. Find the slope and y-intercept for each of the following lines.

   (a) $2y = 4x + 5$ 8(a) _____

   (b) $x - y = 3$ (b) _____

INTRODUCTORY ALGEBRA
CHAPTER 10, TEST F, PAGE 4

9. Write the equation of the line, given the following information.

   (a) The slope is -3/4 and passes through the point (0, 0).

   9(a)_____

   (b) The line passes through the points (6, -1) and (2, 3).

   (b)_____

10. Graph the following inequalities.

   (a) $3x + 4y < 0$

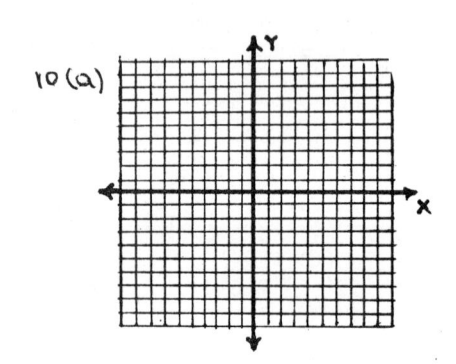

   10(a)

   (b) $x < -2$

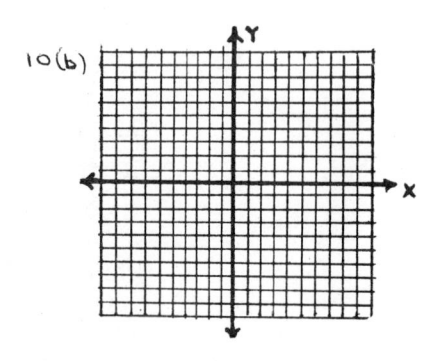

   10(b)

INTRODUCTORY ALGEBRA
CHAPTER 11, TEST A, PAGE 1

1. Solve the following systems of equations graphically. State whether the system is consistent, inconsistent or dependent.

    (a) $y = 3x$
           $y = 4 + x$

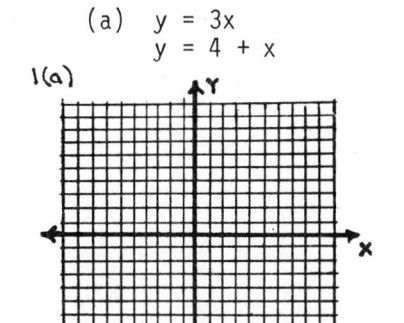

    (b) $3x + y = 9$
           $y + 3x = -9$

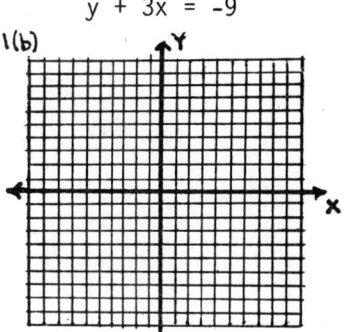

1(a) _____

(b) _____

2. Solve the following systems of equations by the elimination method.

    (a) $y + x = 3$
           $y - x = 1$

    (b) $2x + 5y = 21$
           $6x - y = 15$

2(a) _____

(b) _____

3. Solve the following systems of equations by the substitution method.

    (a) $x = y - 1$
           $3x - 2y = 3$

    (b) $y = \frac{5x}{2}$
           $3x - 2y = 16$

3(a) _____

(b) _____

4. Solve the following systems of equations.

    (a) $2x - y = 6$
           $x + y = 8$

    (b) $3x + 6y = 15$
           $4x + 2y = 10$

    (c) $2x - 2y = 2$
           $3x - 3y = 3$

4(a) _____

(b) _____

(c) _____

INTRODUCTORY ALGEBRA
CHAPTER 11, TEST A, PAGE 2

5. Solve the following problems using two variables.

(a) Kristin Merz invested a sum of money into two different accounts. The first account earned an annual yield of 16% and the second account earned an annual yield of 10%. If the amount of money invested in the first account was $300 more than the sum invested in the second account, and Kristin earned $115.34 from these investments at the end of one year, how much money did she invest in each account?

5(a) _____

_____

(b) Two brands of grass seed sell for $1.90 per pound and $1.20 per pound, respectively. How many pounds of each are needed to produce a fifty pound mixture selling for $1.62 per pound?

(b) _____

_____

INTRODUCTORY ALGEBRA
CHAPTER 11, TEST B, PAGE 1

1. Solve the following systems of equations graphically. State whether the system is consistent, inconsistent or dependent.

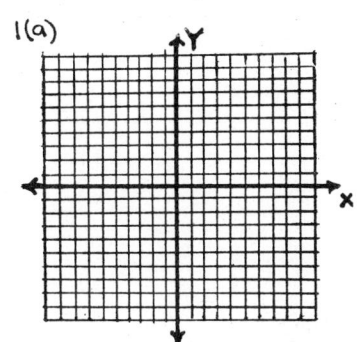

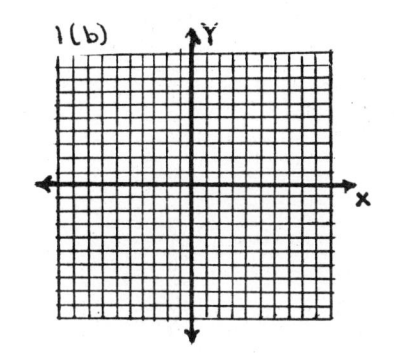

    (a) $3x + y = 6$
           $3x - y = 0$

    (b) $2x + 3y = 4$
           $4x + 6y = 8$

1(a) _____

(b) _____

2. Solve the following systems of equations by the elimination method.

    (a) $3x + y = 1$
           $x - y = 3$

    (b) $3x + y = 9$
           $2x - 5y = 23$

2(a) _____

(b) _____

3. Solve the following systems of equations by the substitution method.

    (a) $x = 5 - 2y$
           $x + 5y = 2$

    (b) $3x - 11y = -2$
           $y = 2x - 5$

3(a) _____

(b) _____

4. Solve the following systems of equations.

    (a) $x + 3y = 4$
           $-x - 6y = 2$

    (b) $5x + 2y = 30$
           $3x - y = -4$

4(a) _____

(b) _____

    (c) $3x - 2y = 9$
           $5x - 3y = 8$

(c) _____

INTRODUCTORY ALGEBRA
CHAPTER 11, TEST B, PAGE 2

5. Solve the following problems using two variables.

(a) Michael rode his bicycle a distance of seventy-two miles. One-half of that distance, he rode against the wind, and the other half, he rode with the wind. It took him three hours against the wind and two hours with the wind to bicycle the seventy-two miles. Find his rate of speed in still air and the rate of speed of the wind.

5(a) _____

_____

(b) A high quality grade top soil can be purchased for $3.60 per "load," while a lower quality grade top soil is $2.00 per "load." If one hundred pounds of a mixture of these two grades of top soil is purchased for $3.00 per "load," how many "loads" of each grade is included in the mix?

(b) _____

_____

INTRODUCTORY ALGEBRA
CHAPTER 11, TEST C, PAGE 1

1. Solve the following systems of equations graphically. State whether the system is consistent, inconsistent or dependent.

    (a) $x + y = 1$
           $x - y = -3$

    (b) $2y = 4x$
           $y = 2x - 5$

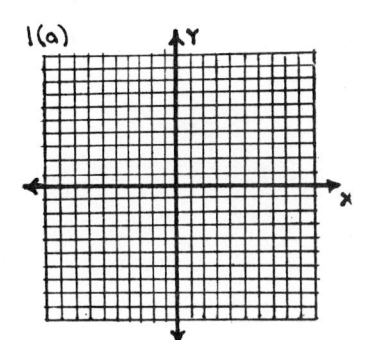

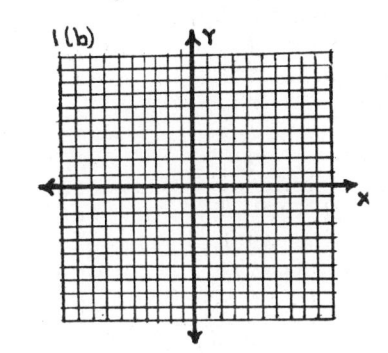

1(a)_____

(b)_____

2. Solve the following systems of equations by the elimination method.

    (a) $x + 2y = 3$
           $x + y = 0$

    (b) $2x + 3y = 6$
           $3x + 5y = 15$

2(a)_____

(b)_____

3. Solve the following systems of equations by the substitution method.

    (a) $x = -2y - 2$
           $4x + 3y = 7$

    (b) $y = \frac{1}{2}x - 2$
           $4x - 3y = 1$

3(a)_____

(b)_____

4. Solve the following systems of equations.

    (a) $5x - 2y = 14$
           $-12x + 2y = -7$

    (b) $6x - 4y = -12$
           $3x + 3y = 9$

    (c) $7x + 3y = -13$
           $3x - 5y = 7$

4(a)_____

(b)_____

(c)_____

INTRODUCTORY ALGEBRA
CHAPTER 11, TEST C, PAGE 2

5. Solve the following problems using two variables.

(a) Robert Nenno purchased twenty-cent and thirty-six cent stamps at the Pittsford Town Post Office. The number of twenty-cent stamps bought was equal to four less than twice the number of thirty-six cent stamps. If he paid a total of $15.92 for the stamps, how many of each kind did he purchase?

5(a)_____

_____

(b) Jim Morrison invested some money into two different accounts. His first investment earned 5% and his second investment earned 8%. He invested three times as much money in the first account, less $2.00, than he did in the second account. If the total amount of money earned at the end of one year from these two investments was $28.65, how much did he invest at each rate?

(b)_____

_____

INTRODUCTORY ALGEBRA
CHAPTER 11, TEST D, PAGE 1

1. Solve the following systems of equations graphically. State whether the system is consistent, inconsistent or dependent.

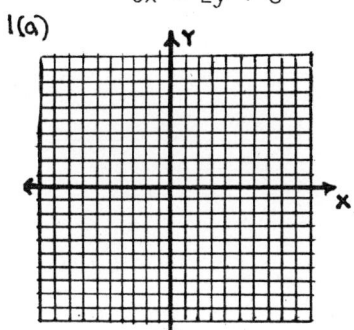

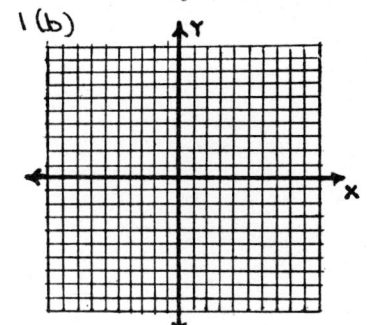

    (a) $3x - y = 4$
          $6x = 2y + 8$

    (b) $2y = 3x$
          $x = 2y + 8$

1(a) _____

(b) _____

2. Solve the following systems of equations by the elimination method.

    (a) $3x - y = 4$
          $x - y = 2$

    (b) $5x - 2y = 20$
          $2x + 3y = 27$

2(a) _____

(b) _____

3. Solve the following systems of equations by the substitution method.

    (a) $x = 2y$
          $2x - 8y = -16$

    (b) $10x + 7y = 19$
          $y = \frac{1}{3}x + \frac{11}{6}$

3(a) _____

(b) _____

4. Solve the following systems of equations.

    (a) $-2x + 3y = -4$
          $2x - 5y = -6$

    (b) $x + 3y = 9$
          $3x - 6y = 12$

    (c) $4x - 2y = 10$
          $3x + 5y = 14$

4(a) _____

(b) _____

(c) _____

INTRODUCTORY ALGEBRA
CHAPTER 11, TEST D, PAGE 2

5. Solve the following problems using two variables.

(a) An airplane traveling with the wind traveled two thousand miles in five hours. On the return trip, howver, it traveled against the wind and completed the trip in eight hours. Find the rate of speed of the airplane in still air and the rate of speed of the wind.

5(a)_____

_____

(b) A game concession operator collected $25.80 in dimes and quarters. If the number of dimes was equal to six more than twice as many quarters, how many dimes and quarters did he collect?

(b)_____

_____

INTRODUCTORY ALGEBRA
CHAPTER 11, TEST E, PAGE 1

1. Solve the following systems of equations graphically. State whether the system is consistent, inconsistent or dependent.

    (a) $y = 3x - 6$
            $2y = 6x - 6$

    (b) $2y = -x - 4$
            $2x - 7 = y$

1(a)_____

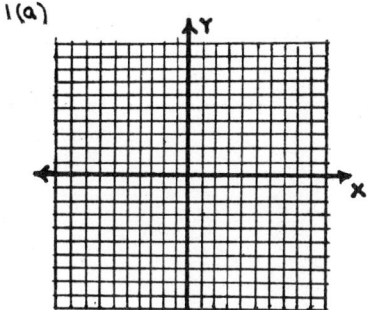

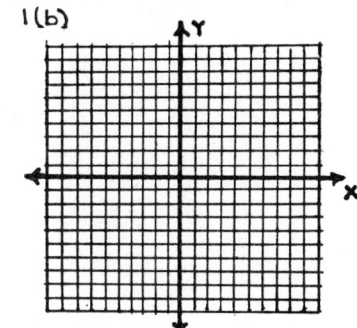

(b)_____

2. Solve the following systems of equations by the elimination method.

    (a) $2x + 3y = 2$
            $5x + 3y = 14$

    (b) $3x + y = 6$
            $x + 3y = 10$

2(a)_____

(b)_____

3. Solve the following systems of equations by the substitution method.

    (a) $y = -2x$
            $6x + 5y = -12$

    (b) $3x - 2y = 11$
            $x = -2y + 9$

3(a)_____

(b)_____

4. Solve the following systems of equations.

    (a) $x - 3y = -27$
            $5x + 3y = 9$

    (b) $2x - 9y = 12$
            $4x - 3y = -6$

    (c) $3x + 8y = -4$
            $5x + 6y = 10/4$

4(a)_____

(b)_____

(c)_____

INTRODUCTORY ALGEBRA
CHAPTER 11, TEST E, PAGE 2

5. Solve the following problems using two variables.

(a) A sum of money was invested, part at 5% and the remaining part at 7 1/2%, and the total amount earned at the end of one year was $42.50. If the amount of money invested at 5% was $100 more than the amount invested at 7 1/2%, how much was invested at each rate?

5(a)_____

_____

(b) Traveling against the current, it took Steve and Adele Hartman four hours to get from the boat launch site to their favorite seaside resturant, a distance of sixty miles. On the return trip, traveling with the current, it took them half as long. Find the rate of speed of their boat in still water and the rate of speed of the current.

(b)_____

_____

INTRODUCTORY ALGEBRA
CHAPTER 11, TEST F, PAGE 1

1. Solve the following systems of equations graphically. State whether the system is consistent, inconsistent or dependent.

    (a) $3y - x = 6$
           $2x = 3y$

    (b) $3y = -2x + 7$
           $2x + 3y - 5 = 0$

1(a) _____

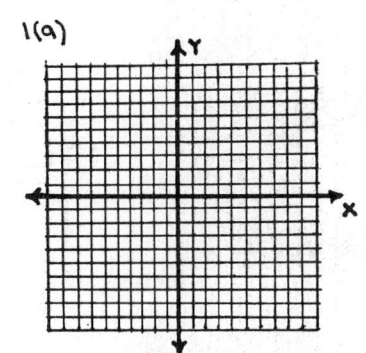

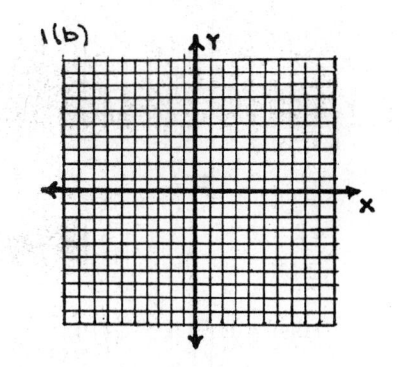

(b) _____

2. Solve the following systems of equations by the elimination method.

    (a) $5y + 4x = 23$
           $y - 4x = -5$

    (b) $5x + 3y = 14$
           $2x + y = 6$

2(a) _____

(b) _____

3. Solve the following systems of equations by the substitution method.

    (a) $y = x + 3$
           $3x + 2y = 26$

    (b) $x = \frac{-3}{7}y + \frac{3}{7}$
           $5x + 6y = 6$

3(a) _____

(b) _____

4. Solve the following systems of equations.

    (a) $x + y = 6$
           $-5x - y = -8$

    (b) $3x + 2y = 7$
           $5x + 8y = 7$

    (c) $5x + 3y = 19$
           $2x + 11y = -12$

4(a) _____

(b) _____

(c) _____

INTRODUCTORY ALGEBRA
CHAPTER 11, TEST F, PAGE 2

5. Solve the following problems using two variables.

(a) At the Elmgrove Farm Market, fresh peaches sell for $0.89 per pound and fresh pears sell for $0.59 per pound. A customer purchased a twenty pound mix of these two fruits for $0.72 per pound. How many pounds of each fruit was included in the mix?

5(a) _____

_____

(b) Larry Gilligan, a blackjack player, cashed in $783 in $2 and $5 chips. The number of $2 chips was equal to three times the number of $5 chips, less ten dollars. How many of each did he have?

(b) _____

_____

INTRODUCTORY ALGEBRA
CHAPTER 12, TEST A, PAGE 1

1. Transform the following equations into standard form.

    (a) $2x(x-4) = 11$ 1(a)_____

    (b) $\dfrac{2(3-x)}{3} = \dfrac{1}{x}$ (b)_____

2. Solve the following equations by factoring. Check your answers in the original equation.

    (a) $x^2 + 5x = 0$ 2(a)_____

    (b) $2x^2 = 32$ (b)_____

    (c) $x^2 + 4x - 21 = 0$ (c)_____

    (d) $2x^2 - 7x + 6 = 0$ (d)_____

3. Solve the following incomplete quadratic equations by using the square root method. Express irrational answers in radical form.

    (a) $3x^2 = 9$ 3(a)_____

    (b) $(x+3)^2 = 4$ (b)_____

4. Solve and check the following equations by completing the square. Answers may be left in radical form.

    (a) $x^2 + 4x - 5 = 0$ 4(a)_____

    (b) $x^2 + 5x + 4 = 0$ (b)_____

    (c) $2x^2 + 4x - 30 = 0$ (c)_____

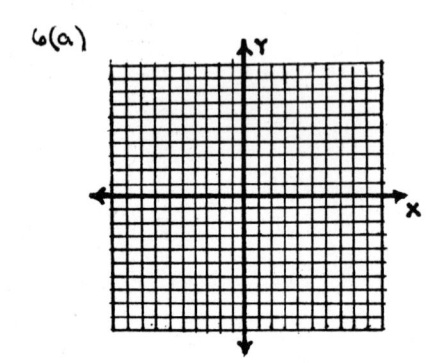

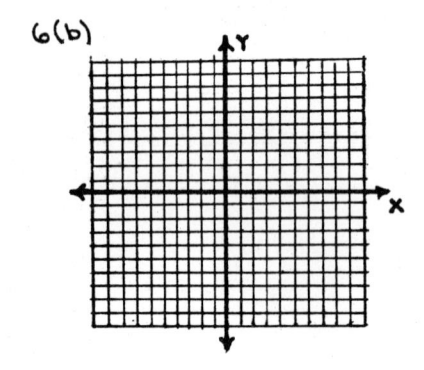

INTRODUCTORY ALGEBRA
CHAPTER 12, TEST A, PAGE 2

5. Solve the following quadratic equations by the quadratic formula. Express irrational answers in radical form.

   (a) $x^2 + 6x + 3 = 0$ 5(a)_____

   (b) $3x^2 - 8x + 4 = 0$ (b)_____

6. Draw the graph of the following quadratic equations. Find the roots from the graph when y = 0.

   (a) $y = x^2 - 6x + 5$ 6(a)_____

   (b) $y = x^2 - 4$ (b)_____

6(a)

6(b)

INTRODUCTORY ALGEBRA
CHAPTER 12, TEST B, PAGE 1

1. Transform the following equations into standard form.

    (a) $4x(x+1) = 5$ 1(a) _____

    (b) $\dfrac{4}{x-2} + \dfrac{8}{x+2} = 1$ (b) _____

2. Solve the following equations by factoring. Check your answers in the original equation.

    (a) $3x^2 = 6x$ 2(a) _____

    (b) $3x^2 - 27 = 0$ (b) _____

    (c) $x^2 - 2x - 8 = 0$ (c) _____

    (d) $4x^2 + 4x + 1 = 0$ (d) _____

3. Solve the following incomplete quadratic equations by using the square root method. Express irrational answers in radical form.

    (a) $2x^2 = 16$ 3(a) _____

    (b) $(x-1)^2 = 9$ (b) _____

4. Solve and check the following equations by completing the square. Answers may be left in radical form.

    (a) $x^2 + 6x + 8 = 0$ 4(a) _____

    (b) $x^2 - 3x - 4 = 0$ (b) _____

    (c) $3x^2 - 6x + 1 = 0$ (c) _____

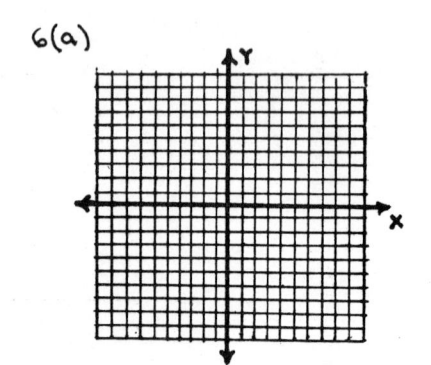

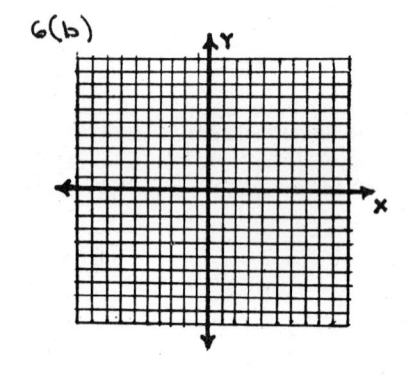

INTRODUCTORY ALGEBRA
CHAPTER 12, TEST B, PAGE 2

5. Solve the following quadratic equations by the quadratic formula. Express irrational answers in radical form.

   (a) $x^2 - 10x - 5 = 0$ 5(a) _____

   (b) $3x^2 + 12x - 7 = 0$ (b) _____

6. Draw the graph of the following quadratic equations. Find the roots from the graph when y = 0.

   (a) $y = x^2 - 4x$ 6(a) _____

   (b) $y = 2x^2 - 2x - 12$ (b) _____

6(a)

6(b)

INTRODUCTORY ALGEBRA
CHAPTER 12, TEST C, PAGE 1

1. Transform the following equations into standard form.

    (a) $y(y+1) = 30$ 1(a)_____

    (b) $\dfrac{x}{3} = \dfrac{9}{x-6}$ (b)_____

2. Solve the following equations by factoring. Check your answers in the original equation.

    (a) $2x^2 - 10x = 0$ 2(a)_____

    (b) $2x^2 = 50$ (b)_____

    (c) $x^2 - 8x + 12 = 0$ (c)_____

    (d) $4y^2 + 21y - 18 = 0$ (d)_____

3. Solve the following incomplete quadratic equations by using the square root method. Express irrational answers in radical form.

    (a) $3x^2 = 24$ 3(a)_____

    (b) $(2x-1)^2 = 8$ (b)_____

4. Solve and check the following equations by completing the square. Answers may be left in radical form.

    (a) $x^2 - 5x - 14 = 0$ 4(a)_____

    (b) $x^2 - 6x + 1 = 0$ (b)_____

    (c) $2x^2 + 6x - 1 = 0$ (c)_____

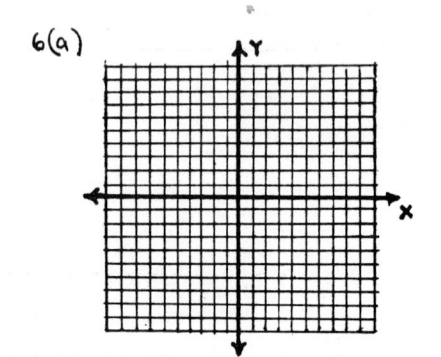

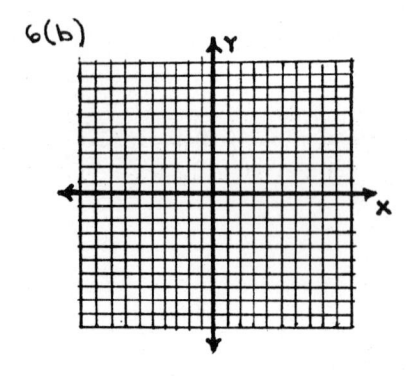

INTRODUCTORY ALGEBRA
CHAPTER 12, TEST C, PAGE 2

5. Solve the following quadratic equations by the quadratic formula. Express irrational answers in radical form.

(a) $x^2 + 9x - 4 = 0$ 5(a) _____

(b) $2x^2 + 5x - 3 = 0$ (b) _____

6. Draw the graph of the following quadratic equations. Find the roots from the graph when y = 0.

(a) $y = x^2 - 9$ 6(a) _____

(b) $y = 3x^2 - 2x - 8$ (b) _____

6(a)

6(b)

INTRODUCTORY ALGEBRA
CHAPTER 12, TEST D, PAGE 1

1. Transform the following equations into standard form.

   (a) $8x(x-2) = 10$ 1(a)_____

   (b) $\dfrac{x^2}{3} + x = \dfrac{1}{6}$ (b)_____

2. Solve the following equations by factoring. Check your answers in the original equation.

   (a) $3x^2 + 15x = 0$ 2(a)_____

   (b) $3y^2 - 75 = 0$ (b)_____

   (c) $x^2 + 10x = -9$ (c)_____

   (d) $3x^2 - 10x + 7 = 0$ (d)_____

3. Solve the following incomplete quadratic equations by using the square root method. Express irrational answers in radical form.

   (a) $x^2 = 24$ 3(a)_____

   (b) $(x+1)^2 = 16$ (b)_____

4. Solve and check the following equations by completing the square. Answers may be left in radical form.

   (a) $x^2 - 9x + 9 = 0$ 4(a)_____

   (b) $x^2 + 6x - 1 = 0$ (b)_____

   (c) $2x^2 - 8x + 3 = 0$ (c)_____

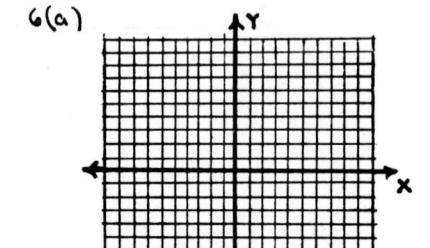

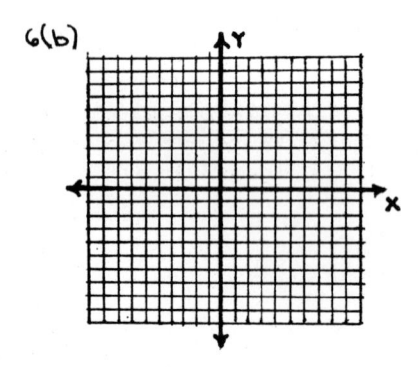

INTRODUCTORY ALGEBRA
CHAPTER 12, TEST D, PAGE 2

5. Solve the following quadratic equations by the quadratic formula. Express irrational answers in radical form.

   (a) $x^2 - 20x + 10 = 0$ 5(a) _____

   (b) $3x^2 + 5x + 1 = 0$ (b) _____

6. Draw the graph of the following quadratic equations. Find the roots from the graph when $y = 0$.

   (a) $y = x^2 - 6x - 16$ 6(a) _____

   (b) $y = x^2 - 4x + 3$ (b) _____

6(a)

6(b)

INTRODUCTORY ALGEBRA
CHAPTER 12, TEST E, PAGE 1

1. Transform the following equations into standard form.

   (a) $(x-1)^2 = 2$ 1(a) _____

   (b) $\dfrac{x^2-2}{3} = \dfrac{5}{6}$ (b) _____

2. Solve the following equations by factoring. Check your answers in the original equation.

   (a) $4x^2 = -12x$ 2(a) _____

   (b) $5x^2 = 45$ (b) _____

   (c) $x^2 - 4x = 5$ (c) _____

   (d) $4x^2 + 16x = 9$ (d) _____

3. Solve the following incomplete quadratic equations by using the square root method. Express irrational answers in radical form.

   (a) $\dfrac{2x}{9} = \dfrac{6}{x}$ 3(a) _____

   (b) $(y+5)^2 = 4$ (b) _____

4. Solve and check the following equations by completing the square. Answers may be left in radical form.

   (a) $x^2 - 10x - 5 = 0$ 4(a) _____

   (b) $x^2 - 3x - 10 = 0$ (b) _____

   (c) $3x^2 - 9x - 7 = 0$ (c) _____

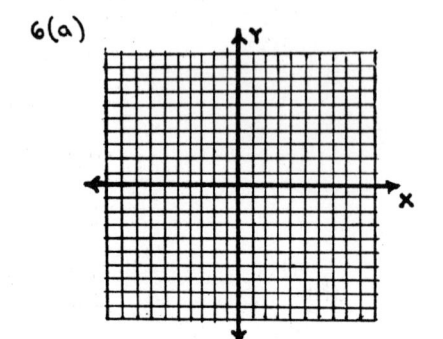

INTRODUCTORY ALGEBRA
CHAPTER 12, TEST E, PAGE 2

5. Solve the following quadratic equations by the quadratic formula. Express irrational answers in radical form.

   (a) $x^2 - x - 5 = 0$ 5(a)_____

   (b) $2x^2 + x - 4 = 0$ (b)_____

6. Draw the graph of the following quadratic equations. Find the roots from the graph when $y = 0$.

   (a) $y = x^2 - 2x - 3$ 6(a)_____

   (b) $y = x^2 - 6x - 7$ (b)_____

6(a)

6(b)

INTRODUCTORY ALGEBRA
CHAPTER 12, TEST F, PAGE 1

1. Transform the following equations into standard form.

    (a) $9(z-5)^2 = 49$ 1(a) _____

    (b) $\dfrac{x-4}{3} = \dfrac{3}{x+4}$ (b) _____

2. Solve the following equations by factoring. Check your answers in the original equation.

    (a) $5x^2 = 5x$ 2(a) _____

    (b) $3x^2 - 12 = 0$ (b) _____

    (c) $x^2 = 8x + 20$ (c) _____

    (d) $2x^2 - 5x = -2$ (d) _____

3. Solve the following incomplete quadratic equations by using the square root method. Express irrational answers in radical form.

    (a) $\dfrac{2x}{8} = \dfrac{8}{x}$ 3(a) _____

    (b) $(z-3)^2 = 12$ (b) _____

4. Solve and check the following equations by completing the square. Answers may be left in radical form.

    (a) $x^2 + 4x - 3 = 0$ 4(a) _____

    (b) $x^2 - 7x - 18 = 0$ (b) _____

    (c) $3x^2 - 15x - 2 = 0$ (c) _____

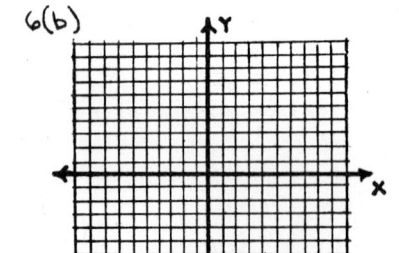

INTRODUCTORY ALGEBRA
CHAPTER 12, TEST F, PAGE 2

5. Solve the following quadratic equations by the quadratic formula. Express irrational answers in radical form.

    (a) $x^2 + 7x - 3 = 0$ 5(a)_____

    (b) $3x^2 - 2x - 6 = 0$ (b)_____

6. Draw the graph of the following quadratic equations. Find the roots from the graph when $y = 0$.

    (a) $y = x^2 + x - 12$ 6(a)_____

    (b) $y = x^2 + 6x + 5$ (b)_____

6(a)

6(b)

FINAL EXAMINATIONS
(FORMS A AND B)

INTRODUCTORY ALGEBRA
FINAL EXAMINATION
(FORM A)

NAME_____DATE_____SCORE_____

In problems 1 - 5, perform the indicated operations.

1. $-4 + 5 \cdot 6 - 7$ 1._____

2. $3 + 8(-3 - 5) \div 4 - (3 - 7)$ 2._____

3. $\dfrac{-2 \cdot 4 + 8}{4 - 6 \cdot 2}$ 3._____

4. $-(6 - 9)^2 + 24 \div 8 + 2 \cdot 5^2$ 4._____

5. $\dfrac{2 \cdot 9 + (3 - 7)^2 - 4}{8 - (5 + 2) - 2^3}$ 5._____

In problems 6 - 9, solve the given equation for x.

6. $2x - 8 = 12$ 6._____

7. $5x + 6 - 2x + 4 = -8$ 7._____

8. $2(3x - 4) - (x + 1) = 2x + 3$ 8._____

9. $x - 5 - 9x + 8 = 3(2x - 7) + 14$ 9._____

10. Solve the formula, $V = \pi r^3 h$ for h.   10._____

In problems 11 - 12, solve the given inequality.
State the solution algebraically, and graph the solution.

11. $3 + x \leq 2x + 4$ 11._____

12. $-3(x - 5) + 6 > 2(x + 4) - 7$ 12._____

FINAL EXAMINATION (FORM A)

13. Choose one of the following word problems and solve it.

   (a) If cashews sell for $4.00 per pound and walnuts sell for $2.75 per pound, how many pounds of each are needed for a 50 pound mixture that will sell for $3.10 per pound?   13(a)_____

   (b) Tim invested part of $10,000 at 8% and the remaining part at 6%. If the total annual income from these investments is $720, how much was invested at each rate?   (b)_____

   (c) Two cars that are 375 miles apart begin traveling towards each other at the same time. The first car is traveling at a rate of speed 15 mph slower than the second car. If the cars pass each other in 3 hours, find the rate of speed of each car.   (c)_____

In problems 14 - 18, perform the indicated operations. Simplify your answers.

14. $(2x^3 - 5x^2 + 4) - (-2x^3 + 3x^2 - 2x + 9)$ 14._____

15. $(2x + 4)(3x - 5)$ 15._____

16. $(x + 6)(3x^2 - 2x + 8)$ 16._____

17. $\dfrac{18x^2y - 9x^6y^4 + 24xy^2z}{3x^4y^2}$ 17._____

18. $\dfrac{x^3 + x^2 - 34x + 20}{x - 5}$ 18._____

FINAL EXAMINATION (FORM A)   NAME_____

In problems 19 - 23, factor the given expresession completely.

19. $5ab - 10ab^2 + 15a^2b$ 19._____

20. $x^2 + 2x - 24$ 20._____

21. $2x^2 - 7x - 15$ 21._____

22. $6x^3 + x^2 - 12x$ 22._____

23. $4x^2 - 25y^2$ 23._____

In problems 24 - 28, simplify the given fractions.

24. $\dfrac{10}{x^2 - x - 6} \cdot \dfrac{x^2 - 5x - 14}{25}$ 24._____

25. $\dfrac{x^2 - 1}{(x - 3)} \div \dfrac{x^2 + x}{(3 - x)}$ 25._____

26. $\dfrac{5}{x} + \dfrac{x + 3}{2x}$ 26._____

27. $\dfrac{x + 2}{x^2 - 9} - \dfrac{x - 4}{x^2 + 7x + 12}$ 27._____

28. $\dfrac{1 - \dfrac{1}{x}}{2 - \dfrac{2}{x}}$ 28._____

FINAL EXAMINATION (FORM A)

In problems 29 - 31, solve the given fractional equations for x.

29. $\dfrac{2}{x} + \dfrac{1}{8} = \dfrac{3}{8}$

29._____

30. $\dfrac{2}{3} - \dfrac{1}{5x} = 7$

30._____

31. $\dfrac{3}{x+4} + \dfrac{1}{8} = \dfrac{5}{8}$

31._____

In problems 32 - 36, simplify the given radical expressions.

32. $\sqrt{72}$

32._____

33. $2\sqrt{3} - 4\sqrt{3}$

33._____

34. $\sqrt{8} + 9\sqrt{18} - 6\sqrt{50}$

34._____

35. $\sqrt{\dfrac{12}{25}}$

35._____

36. $\sqrt{\dfrac{27}{40}}$

36._____

In problems 37 - 39, graph the given equations and inequality.

37. $2x - 3y = 4$ 38. $3x + 2y = 4$ 39. $2x - y < 8$

37.

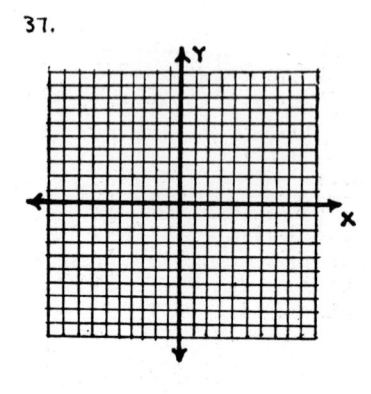

38.

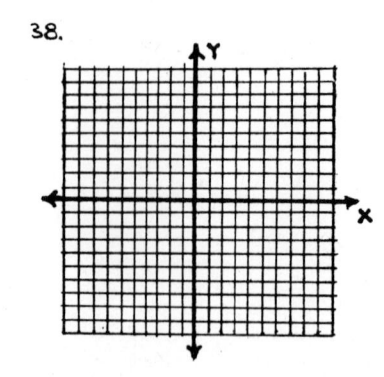

39.

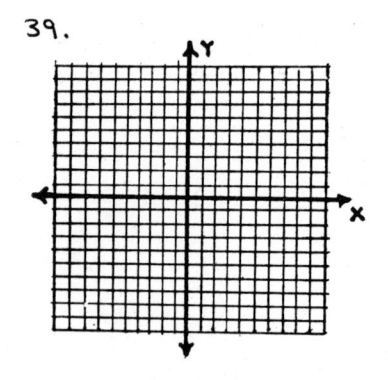

FINAL EXAMINATION (FORM A)   NAME_____

40. Write the equation of the line that passes through the points (0, 0) and (2, 5).   40._____

41. Write the equation of the line that passes through the point (1, 4) and has a slope of 3.   41._____

In problems 42 - 44, solve the given systems of equations.

42. $2x - 3y = 6$
    $3x + 3y = 4$ 42._____

43. $x - 5y = -3$
    $4x + 2y = 54$ 43._____

44. $3x + 5y = 2$
    $2x + 3y = -7$ 44._____

In problems 45 - 49, solve the given quadratic equations for x.

45. $2x^2 - 4x = 0$ 45._____

46. $x^2 - 9 = 0$ 46._____

47. $x^2 - 2x - 3 = 0$ 47._____

48. $x^2 + 2x = 1$ 48._____

49. $3x^2 = 1 - 2x$ 49._____

50. Graph the quadratic equation, $y = x^2 + 2x - 3$, on the given set of axes.

50.

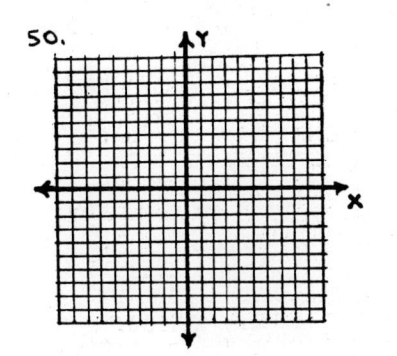

INTRODUCTORY ALGEBRA
FINAL EXAMINATION
(FORM B)

NAME_____DATE_____SCORE_____

In problems 1 - 5, perform the indicated operations.

1. $-2 + 8 \cdot 3 - 5$

1._____

2. $4 + 7(-8 + 5) \div 7 - (7 - 9)$

2._____

3. $\dfrac{4 \cdot (-3) - 7}{5 - 8 \cdot 3}$

3._____

4. $(7 - 8)^2 - 15 \div (-3) + 4 \cdot (-3)^2$

4._____

5. $\dfrac{3 \cdot 6 + (4 - 7)^2 - 3}{6 - (2 - 8) + 4^2}$

5._____

In problems 6 - 9, solve the given equation for x.

6. $3x - 19 = 12$

6._____

7. $2x + 7 - 4x + 6 = -3$

7._____

8. $3(2x - 7) - (2x - 3) = x - 5$

8._____

9. $2x + 4 - 6x - 3 = 2(7 - 3x) + 11$

9._____

10. Solve the formula, $F = \dfrac{9}{5}C + 32$ for C

10._____

In problems 11 - 12, solve the given inequality.
State the solution algebraically, and graph the solution.

11. $x - 4 > 3 - 2x$

11._____

12. $-2(x + 3) - 6 \leq 3(4 - x) + 9$

12._____

FINAL EXAMINATION (FORM B)

13. Choose one of the following word problems and solve it.

    (a) Two cars leave from the same place and the same time and travel in opposite directions. The first car's rate of speed is 40 mph, and the second car's rate of speed is 28 mph. In how many hours will the two cars be 187 miles apart?

13(a) _____

    (b) A change vending machine contains $20.25 in quarters, dimes and nickels. If the number of quarters is 17 more than the number of dimes, and there are 8 more dimes than nickels, how many of each coin are contained in the machine?

(b) _____

    (c) A pair of pants cost $32 more than a shirt. If 2 pairs of pants and 3 shirts cost $229, find the cost of each.

(c) _____

In problems 14 - 18, perform the indicated operations. Simplify your answers.

14. $(3x^3 - 6x^2 - 2x + 4) - (5x^3 - 2x^2 + 4x - 6)$

14. _____

15. $(2x - 3)(5x + 6)$

15. _____

16. $(x - 3)(2x^2 - 7x + 9)$

16. _____

17. $\dfrac{5ab - 15a^4b^2c^3 + 20a^2b^2c^3}{5a^2b^2c}$

17. _____

18. $\dfrac{2x^3 - 2x^2 - 17x + 15}{(x - 3)}$

18. _____

FINAL EXAMINATION (FORM B)   NAME_____

In problems 19 - 23, factor the given expresession completely.

19. $4x^2y^2 - 8xy^3 - 16x^5y^3$ 19._____

20. $x^2 + x - 20$ 20._____

21. $3x^2 - 19x - 40$ 21._____

22. $6x^3 + x^2 - 2x$ 22._____

23. $16x^2 - 64y^2$ 23._____

In problems 24 - 28, simplify the given fractions.

24. $\dfrac{8}{x^2 + 5x + 6} \cdot \dfrac{x^2 + 4x + 4}{20}$ 24._____

25. $\dfrac{(5 - x)}{2(x + y)} \div \dfrac{(x - 5)}{4x^2 - 4y^2}$ 25._____

26. $\dfrac{3}{x} + \dfrac{2x - 5}{3x}$ 26._____

27. $\dfrac{x - 7}{x^2 + 7x + 10} - \dfrac{x + 3}{x^2 - 25}$ 27._____

28. $\dfrac{2 + \dfrac{3}{x}}{5 - \dfrac{4}{x}}$ 28._____

FINAL EXAMINATION (FORM B)

In problems 29 - 31, solve the given fractional equations for x.

29. $\dfrac{5}{x} + \dfrac{1}{7} = \dfrac{5}{7}$ 29. _____

30. $\dfrac{3}{4} - \dfrac{6}{3x} = 9$ 30. _____

31. $\dfrac{2x}{x-2} + \dfrac{3}{7} = \dfrac{9}{x-2}$ 31. _____

In problems 32 - 36, simplify the given radical expressions.

32. $-\sqrt{108}$ 32. _____

33. $5\sqrt{7} + 2\sqrt{7}$ 33. _____

34. $\sqrt{20} - 2\sqrt{80} + 3\sqrt{45}$ 34. _____

35. $\sqrt{\dfrac{63}{16}}$ 35. _____

36. $\sqrt{\dfrac{28}{75}}$ 36. _____

In problems 37 - 39, graph the given equations and inequality.

37. $2x + y = 9$ 38. $5x - 2y = 10$ 39. $3x + 2y > 9$

37.

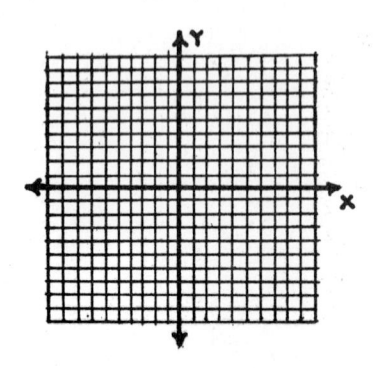

38.

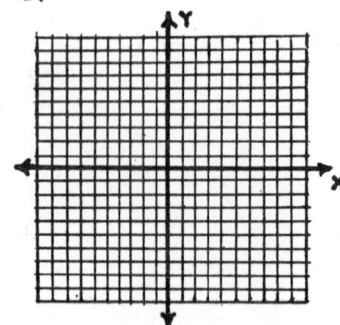

39.

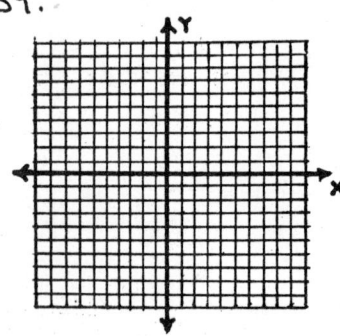

FINAL EXAMINATION (FORM B)   NAME_____

40. Write the equation of the line that passes through the point (0, 5) and has a slope of 2.

    40._____

41. Write the equation of the line that passes through the points (5, 0) and (0, -3).

    41._____

In problems 42 - 44, solve the given systems of equations.

42. $x + 2y = 3$
    $-x - 3y = 8$

    42._____

43. $3x + 4y = -5$
    $x + 2y = 9$

    43._____

44. $5x + 8y = 8$
    $3x - 3y = -3$

    44._____

In problems 45 - 49, solve the given quadratic equations for x.

45. $3x^2 - 12x = 0$

    45._____

46. $x^2 - 36 = 0$

    46._____

47. $x^2 - 2x - 15 = 0$

    47._____

48. $x^2 = 2x + 1$

    48._____

49. $6x^2 = 2 - x$

    49._____

50. Graph the quadratic equation, $y = x^2 - 2x - 3$, on the given set of axes.

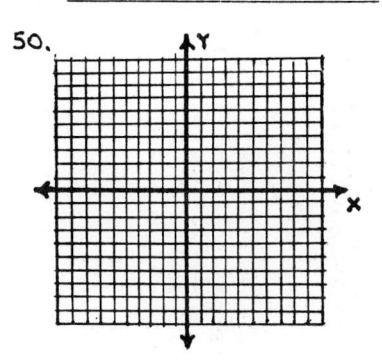

    50.

CHAPTER TESTS
ANSWERS

CHAPTER 1 TEST A

1. $1\frac{1}{2}$
2. $5/10$
3. 0.4
4. $3/2$
5. $2\frac{2}{3}$
6a. $0 < 4$
b. $4 > 3$
c. $0 = 0$
7a. $0.9 < 1.1$
b. $0.5 > 0.05$
c. $3.2 = 3.2$
8a. $\frac{1}{2} = \frac{3}{6}$
b. $\frac{1}{4} < \frac{1}{3}$
c. $\frac{5}{9} > \frac{3}{9}$
9a. 599
b. 609
10a. 0.365
b. 3.25
11a. $4/7$
b. $7/6$
12a. $4\frac{1}{2}$
b. $5\frac{1}{2}$
13a. 127
b. 51
14a. 12.2
b. 2.3
15a. $1/4$
b. $1/3$
16a. $6\frac{1}{5}$
b. $12\frac{3}{8}$
17a. 381
b. 15.450
18a. 0.72
b. 0.245
19a. $9/40$
b. $8/15$
20a. $4\frac{1}{2}$
b. $1\frac{7}{8}$
21a. 13
b. 32
22a. 4.2
b. 6.4
23a. $5/7$
b. $2/3$
24a. $1\frac{13}{32}$
b. $13\frac{1}{3}$
25a. $7 \times^3$
b. $2^2 \cdot 3^3$
26a. 17
b. 11
27a. 11
b. 21

CHAPTER 1 TEST B

1. 1
2. $5/100$
3. 0.25
4. $8/3$
5. $2\frac{1}{4}$
6a. $7 > 0$
b. $3 = 3$
c. $2 < 6$
7a. $0.12 > 0.06$
b. $2.99 < 5.1$
c. $0.125 = 0.125$
8a. $\frac{1}{2} < \frac{3}{4}$
b. $\frac{3}{8} = \frac{9}{24}$
c. $\frac{7}{9} > \frac{3}{4}$
9a. 756
b. 523
10a. 0.179
b. 1.421
11a. $11/10$
b. $3/3$ OR 1
12a. 6
b. $3\frac{5}{6}$
13a. 1424
b. 28
14a. 9.3
b. 9.75
15a. 3
b. $3/10$
16a. $1\frac{1}{4}$
b. $11\frac{2}{5}$
17a. 1495
b. $18,450$
18a. 0.12
b. 0.052
19a. $9/65$
b. $1/3$
20a. 12
b. 4
21a. 16
b. 70
22a. 0.23
b. 0.23
23a. $4/3$
b. $4/7$
24a. $2\frac{1}{82}$
b. $2/63$
25a. $x^2 \cdot 3^2$
b. 4^5
26a. 17
b. 36
27a. 7
b. 20

CHAPTER 1
TEST C

1. 3
2. $25/100$
3. 0.5
4. $29/8$
5. $1\frac{1}{2}$
6a. $4 = 4$
b. $1 < 9$
c. $5 > 2$
7a. $0 > 0.001$
b. $3.12 = 3.1200$
c. $0.4 > 0.125$
8a. $\frac{6}{9} = \frac{2}{3}$
b. $\frac{1}{8} < \frac{3}{8}$
c. $2\frac{1}{2} > \frac{2}{3}$
9a. 487
b. 7898
10a. 2.999
b. 6
11a. $3/5$
b. $5/4$
12a. $14\frac{1}{4}$
b. $6\frac{1}{12}$
13a. 180
b. 3876
14a. 4.8
b. 1.125
15a. $1/3$
b. $1/8$
16a. $2\frac{3}{5}$
b. $3\frac{1}{4}$
17a. 742
b. 5995
18a. 0.425
b. 7.75
19a. $1/12$
b. $2/3$
20a. $7\frac{21}{32}$
b. $9\frac{3}{5}$
21a. 40
b. 16
22a. 8.03
b. 51
23a. $2\frac{2}{3}$
b. $3/5$
24a. $20/27$
b. 24
25a. $7^3 \cdot x^2$
b. $y^3 \cdot 2^1$
26a. 6
b. 200
27a. 14
b. 5

CHAPTER 1
TEST D

1. $1/2$
2. $7/10$
3. 0.125
4. $13/4$
5. 1
6a. $7 > 6$
b. $0 < 9$
c. $2 = 2$
7a. $0.4 = 0.40$
b. $2.1 > 0.9$
c. $0.135 < 0.8$
8a. $2\frac{1}{2} > 2\frac{1}{4}$
b. $\frac{1}{2} = \frac{4}{8}$
c. $\frac{2}{3} < \frac{3}{4}$
9a. 778
b. 2115
10a. 0.542
b. 4.211
11a. $7/6$
b. $1/2$
12a. 6
b. $6\frac{1}{4}$
13a. 140
b. 2976
14a. 5.75
b. 11.16
15a. $1/2$
b. $1/6$
16a. $8\frac{2}{3}$
b. $3\frac{11}{12}$
17a. 2920
b. 11,580
18a. 0.0203
b. 0.0006
19a. $25/42$
b. $2/9$
20a. 20
b. 6
21a. 104
b. 43
22a. 8.2
b. 26
23a. $20/9$
b. $10/9$
24a. $1\frac{11}{16}$
b. 18
25a. $x^2 \cdot 2^1$
b. $2^3 \cdot x^2 \cdot y^1$
26a. 18
b. 72
27a. 7
b. 12

CHAPTER 1
TEST E

1. $3\frac{1}{2}$
2. $\frac{50}{100}$
3. 0.75
4. $\frac{23}{8}$
5. $2\frac{1}{5}$
6a. 2 < 4
b. 5 = 5
c. 1 > 0
7a. 0.9 > 0.89
b. 0.09 < 0.9
c. 3.001 = 3.001
8a. $1 < \frac{10}{9}$
b. $\frac{5}{7} < \frac{6}{8}$
c. $\frac{1}{3} = \frac{3}{9}$
9a. 353
b. 5555
10a. 1.495
b. 3.5
11a. $\frac{1}{2}$
b. $\frac{11}{15}$
12a. $5\frac{3}{5}$
b. $4\frac{9}{10}$
13a. 1411
b. 1229
14a. 0.013
b. 12.87
15a. $\frac{2}{5}$
b. $\frac{4}{15}$
16a. $6\frac{3}{5}$
b. $1\frac{1}{5}$
17a. 4320
b. 32,614
18a. 57.33
b. 1.26
19a. $\frac{3}{20}$
b. $\frac{1}{9}$
20a. $1\frac{17}{18}$
b. $3\frac{6}{7}$
21a. 91
b. 25
22a. 6.03
b. 36.7
23a. $\frac{5}{9}$
b. $\frac{3}{5}$
24a. $\frac{1}{12}$
b. $\frac{5}{12}$
25a. $x^2 \cdot y^2$
b. $x^2 \cdot y^3$
26a. 10
b. 72
27a. 8
b. 3

CHAPTER 1
TEST F

1. 2
2. $\frac{3}{10}$
3. 0.8
4. $\frac{19}{4}$
5. 5
6a. 10 > 8
b. 1 = 1
c. 3 < 7
7a. 0.16 < 0.167
b. 0.25 = 0.25
c. 3.5 > 3.14
8a. $\frac{9}{10} < 1\frac{1}{2}$
b. $\frac{2}{8} = \frac{1}{4}$
c. $1\frac{1}{9} < \frac{9}{8}$
9a. 262
b. 10,158
10a. 0.485
b. 1.08
11a. $\frac{8}{9}$
b. $\frac{2}{3}$
12a. 12
b. $11\frac{1}{5}$
13a. 1393
b. 192
14a. 98.17
b. 3.18
15a. $\frac{2}{9}$
b. $\frac{1}{12}$
16a. $4\frac{1}{6}$
b. $5\frac{9}{16}$
17a. 1455
b. 17,920
18a. 0.12
b. 7.8
19a. $\frac{3}{8}$
b. $\frac{1}{6}$
20a. 12
b. 39
21a. 979
b. 4
22a. 18.9
b. 8.5
23a. 1
b. $\frac{9}{5}$
24a. $\frac{17}{22}$
b. $\frac{11}{16}$
25a. $x^1 \cdot y^3$
b. $3^2 \cdot x^3 \cdot z^1$
26a. 1
b. 100
27a. 14
b. 10

CHAPTER 2 TEST A	CHAPTER 2 TEST B	CHAPTER 2 TEST C	CHAPTER 2 TEST D
1a. +15	1a. +14	1a. +12	1a. +13
b. +10	b. +26	b. -13	b. +6
2. (c)	2. (b)	2. (a)	2. (b)
3a. -6	3a. +15	3a. +4	3a. -21
b. +3	b. -10	b. -4	b. -18
c. +5 2/3	c. -10 4/5	c. -8 2/7	c. -1 1/8
4a. +18	4a. +20	4a. -16	4a. -15
b. -28	b. -13	b. -65	b. -8
5a. -30	5a. -28	5a. -40	5a. +56
b. -6	b. -6/5	b. -5/3	b. +5/4
c. +7/24	c. +5/11	c. -88/27	c. +1/3
d. +12	d. -7	d. -12	d. -56
6a. -2	6a. +10	6a. -9	6a. -9
b. +18	b. -80	b. -125	b. -72
7a. -3	7a. +5	7a. -2	7a. -5
b. -45	b. +16	b. -16	b. +3
c. +4	c. -16	c. -4	c. +5
8a. -34	8a. -24	8a. +16	8a. -4
b. +2	b. -4	b. -72	b. -9
c. -3	c. -11/3	c. +5/3	c. 4/5
9a. COMMUTATIVE PROPERTY (+)	9a. ASSOCIATIVE PROPERTY (+)	9a. IDENTITY PROPERTY (×)	9a. INVERSE PROPERTY (+)
b. MULTIPLICATION PROPERTY "0"	b. IDENTITY PROPERTY (+)	b. ZERO PROPERTY (×)	b. DISTRIBUTIVE PROPERTY
c. DISTRIBUTIVE PROPERTY	c. DISTRIBUTIVE PROPERTY	c. COMMUTATIVE PROPERTY (×)	c. COMMUTATIVE PROPERTY (×)

CHAPTER 2 TEST E	CHAPTER 2 TEST F	CHAPTER 3 TEST A	CHAPTER 3 TEST B
1a. $+11$	1a. $+11$	1a. 2 TERMS;	1a. 2 TERMS;
b. -25	b. -40	2a AND 3b;	3a AND b;
2. (d)	2. (c)	2 AND 3	3 AND 1
3a. -19	3a. -32	b. 1 TERM;	b. 1 TERM;
b. $+8$	b. -25	$2(a+b)$;	$x+3y$;
c. $-5\frac{1}{2}$	c. $+6\frac{4}{9}$	2	1
4a. -7	4a. $+11$	2a. x^2	2a. $x^2 y^2$
b. $+14$	b. $+12$	b. $3x^2 y^2$	b. $-3xy^3$
5a. $+36$	5a. $+50$	3a. $x \cdot x$	3a. $(x)(x)(y)$
b. $+25/4$	b. $-2/9$	b. $(-2)(-2)(-2)(x)(x)(x)$	b. $(-2)(-2)(-2)(-2)(x)(x)(x)(x)$
c. $-2/7$	c. $-1/3$	4a. $+20$	4a. $+38$
d. -25	d. -60	b. -4	b. $+12$
6a. $+2/5$	6a. $+2$	5a. $2x$	5a. $2x$
b. $+1$	b. -16	b. $3x-5y$	b. $-2x-7y$
7a. $+4$	7a. -7	6a. $3x+2y$	6a. $x+y$
b. $-3/2$	b. $-20/27$	b. $5x-2$	b. $5a+b$
c. -9	c. -21		
8a. $+12$	8a. -9		
b. -14	b. -86		
c. -19	c. $-9/20$		
9a. COMMUTATIVE PROPERTY ($\times$)	9a. COMMUTATIVE PROPERTY ($+$)		
b. DISTRIBUTIVE PROPERTY	b. INVERSE PROPERTY ($+$)		
c. ASSOCIATIVE PROPERTY ($\times$)	c. IDENTITY PROPERTY ($+$)		

CHAPTER 3 TEST C

1a. 1 TERM; $6xy$; 6
b. 1 TERM; $2(x+y)/3$; $2/3$
2a. xy^2
b. $2^2 x^2 yz$
3a. $(2)(x)(x)(x)$
b. $(-3)(-3)(x)(x)(x)(x)$
4a. -4
b. $-2/5$
5a. $+11y$
b. $3x+z$
6a. $5x+y$
b. $-x+10$

CHAPTER 3 TEST D

1a. 1 TERM; $-13xyz$; -13
b. 2 TERMS; $\frac{x}{2}$ AND $\frac{x}{3}$; $\frac{1}{2}$ AND $\frac{1}{3}$
2a. $x^2 y$
b. $(2)(3)(x^3)$
3a. $(-2)(y)(y)$
b. $(2)(2)(2)(x)(x)(x)(y)(y)(y)$
4a. -30
b. -10
5a. $4ab$
b. $20a - 11ab$
6a. $8y - 2x$
b. $-x - 7y$

CHAPTER 3 TEST E

1a. 1 TERM; $\frac{x}{3}$; $\frac{1}{3}$
b. 2 TERMS; $2(x+y)$ AND -4; 2 AND -4
2a. $x^2 y^2 z^2$
b. $(3)(5)(x^3)(y^2)$
3a. $(3)(x)(x)(y)(y)$
b. $(3)(3)(x)(x)(x)(x)(y)(y)(y)(y)$
4a. $+48$
b. $+6$
5a. 0
b. $4x - x^2$
6a. $-6x + y$
b. $-7x - 3y$

CHAPTER 3 TEST F

1a. 2 TERMS; $9x$ AND by; a AND b.
b. 1 TERM; $\frac{x-3}{2}$; $\frac{1}{2}$
2a. $x^5 y$
b. $(-2)(-3)(5)(x^4)$
3a. $(-4)(x)(x)(y)(y)(y)$
b. $(-1)(-1)(-1)(-1)(x)(x)(x)(x)(x)(x)(x)(y)(y)(y)(y)$
4a. $+14$
b. $+14$
5a. $-2x^2$
b. $3x + 4x^2$
6a. $4x - 3y$
b. $10a + 10b$

CHAPTER 4 TEST A	CHAPTER 4 TEST B	CHAPTER 4 TEST C	CHAPTER 4 TEST D
1a. $x=3$	1a. $x=8$	1a. $x=4$	1a. $x=5$
b. $x=6$	b. $x=50$	b. $x=6$	b. $x=4$
c. $x=1$	c. $x=5$	c. $x=2$	c. $x=0$
d. $x=9$	d. $x=3$	d. $x=4$	d. $x=6$
2a. $x=3$	2a. $x=36$	2a. $x=120$	2a. $x=7$
b. $x=18$	b. $x=2$	b. $x=23/5$	b. $x=21$
3a. $x=4$	3a. $x=3$	3a. $x=1$	3a. $x=14$
b. $x=-3$	b. $x=4$	b. $x=6$	b. $x=-12$
4a. $x=7/2$	4a. $x=2$	4a. $x=6$	4a. $x=7$
b. $x=4$	b. $x=9$	b. $x=5$	b. $x=20$
5a. $x=0$	5a. $x=1$	5a. $x=-2$	5a. $x=-13$
b. $x=-1$	b. $x=-4$	b. $x=1$	b. $x=6$
6a. $r = C/2\pi$	6a. $b = A/h$	6a. $d = C/\pi$	6a. $h = 2A/b$
b. $\ell = \frac{P-2W}{2}$	b. $F = \frac{9}{5}\left(K + \frac{160}{9}\right)$	b. $y = \frac{6R}{x}$	b. $r = \frac{A-p\ell}{pt}$
7a. $x>2$	7a. $x>5$	7a. $x \leq 6$	7a. $x \geq 0$
b. $x \geq 4$	b. $x \geq -3$	b. $x < -3$	b. $x > -4$

CHAPTER 4 TEST E

1a. $x = 1$
b. $x = 1$
c. $x = 3$
d. $x = 11$
2a. $x = 3$
b. $x = 45$
3a. $x = 20$
b. $x = 11$
4a. $x = 3/4$
b. $x = 6$
5a. $x = 2$
b. $x = 5$
6a. $R = \frac{D}{T}$
b. $C = \frac{5}{9}(F-32)$
7a. $x > -1$
b. $x \geq -3$

CHAPTER 4 TEST F

1a. $x = 1/2$
b. $x = 9$
c. $x = 3$
d. $x = 6$
2a. $x = 50$
b. $x = 19/16$
3a. $x = 0$
b. $x = -3$
4a. $x = 1/5$
b. $x = 5/8$
5a. $x = 8$
b. $x = -1$
6a. $W = \frac{V}{\ell b}$
b. $C = \frac{2A - hb}{h}$
7a. $x > 1$
b. $x \leq -1/2$

CHAPTER 5 TEST A

1. -2
2. BENNY: 45
   MARIA: 23
3. 20 MILES
4. 1.8 lb. Almonds
   1.2 lb. TRAIL MIX
5. PIZZA: $24,000
   BIKE: $60,000
6. DON: 35 MPH
   NANCY: 23 MPH
7. $2.04 @ lb.
8. $7500 @ 7%
   $2500 @ 15%

CHAPTER 5 TEST B

1. -14
2. 16 AND 4
3. 6 lb. sunflower
   4 lb. CASHEWS
4. Michael: 24 MPH
   JANE: 15 MPH
5. $6600 @ 12%
   $5100 @ 8%
6. 250 NICKELS
   286 DIMES
   310 quarters
7. $12,500 @ 18%
   $7,500 @ 10%
8. 27 MINUTES

CHAPTER 5 TEST C	CHAPTER 5 TEST D	CHAPTER 5 TEST E	CHAPTER 5 TEST F
1. −1	1. 2	1. 25	1. −4
2. SHIRT: $24 PANTS: $39	2. 21 AND 34	2. RON: 24 GEORGE: 18	2. SOCKS: $2.50 SHOES: $25
3. 3 HR. BICYCLING 2 HR. RUNNING	3. 6 lb. MUTSU 4 lb. CORTLAND	3. $580 @ 12% $1164 @ 15%	3. APPLE +: 40 qt. OCEAN MIST: 60 qt.
4. $1350 @ 12% $3000 @ 15%	4. $30,000 @ 12% $20,000 @ 8½%	4. $22.30 @ DOZEN	4. $92 @ 5% $392 @ 8%
5. 300 ADVANCE 550 DOOR	5. TO AIRPORT: 1¼ HR. FROM AIRPORT: 1¾ HR.	5. VEHICLE #1: 90 MPH VEHICLE #2: 60 MPH	5. 30 MINUTES
6. 8.6 HOURS	6. 4.8 MILES	6. $1.87 @ lb.	6. 1492 @ $12.50 4476 @ $8.50 2980 @ $15.00
7. $4800 @ 15% $3200 @ 12½%	7. 75 lb. PORK	7. $1125 @ 13% $1875 @ 5%	7. 6 HR. @ 45 MPH 2 HR @ 52 MPH
8. $2.68 @ lb.	8. $600 @ 9% $400 @ 6½%	8. 60 MILES	8. $1800 @ 10% $700 @ 5.5%

CHAPTER 6 TEST A	CHAPTER 6 TEST B	CHAPTER 6 TEST C	CHAPTER 6 TEST D
1a. $6y$	1a. $-6x$	1a. $7x^2$	1a. $11x^3$
b. $4x$	b. 0	b. $12y$	b. $-7x$
c. $11x^2$	c. $-3(a+b)$	c. $3x$	c. $-3x$
2a. $7+4x$	2a. $3x+5y$	2a. $4r^2-7$	2a. $14x^2+2xy-1$
b. $-5x^2+3x+5$	b. $6-2x$	b. $5w^2$	b. $2x^2+4$
c. $8x^2-x-11$	c. $9x^2+4y^2$	c. $-5x^2+12x+40$	c. 0
3a. x	3a. $-8x$	3a. $-3x^2$	3a. $3ab$
b. $-5x$	b. $13x^2$	b. $6ab$	b. $3x^2$
4a. $x-3y$	4a. $5ab+4a+2$	4a. $-1+4x^2y-x-y$	4a. x^2+x
b. $-2x$	b. $x-1$	b. $5x+1$	b. $4x^2-2x$
c. $-xy$	c. $5x+18y$	c. $-5x^2+7xy-5y^2$	c. $2x^2+12xy-5y^2$
5a. x^5	5a. a^5	5a. y^6	5a. x^3
b. a^6	b. y^8	b. z^6	b. x^{16}
c. a^2b^2	c. x^2y^4	c. x^6y^6	c. x^3y^9
d. $-8x^6y^3$	d. $27x^3y^3$	d. $9x^6y^2$	d. $25x^2y^4$
6a. $20a^5$	6a. $-9x^6$	6a. $30x^{10}$	6a. $-9x^5$
b. $28x^2+84x$	b. $20x^3+16x^2-32x$	b. $-x^4-3x^3+7x^2$	b. $90x^2-150x$
c. $x^2-5x-36$	c. $x^2+9x+14$	c. $2x^2+x-15$	c. $2a^2+7ab-15b^2$
d. $8x^3-60x^2+150x-125$	d. $-2x^3+3x+1$	d. $2x^3-15x^2-14x-24$	d. $x^3-8x^2+11x+20$
7a. x^3	7a. y^4	7a. x	7a. x^2
b. y^3	b. y	b. x^4	b. 1
c. $1/x$	c. $1/y^5$	c. $1/y^3$	c. $1/x^3$
d. $7/x$	d. $1/y^2$	d. x^2/z	d. y/x
8a. $-7ab$	8a. $1/3y^3$	8a. $8xy$	8a. $-8x^2/y^2$
b. $x+2$	b. $\frac{1}{yz}+\frac{1}{xz}+\frac{1}{xy}$	b. $-4x^2+2x+1$	b. $-2x^2y^2+xy-3$
c. $3x^2-5x+2$	c. $x+5$	c. $x+2$	c. $x-3$

CHAPTER 6 TEST E

1a. $7y$
b. $5x$
c. $-9x$
2a. $2x^3 + 2x^2 - 4x$
b. $8x - 8$
c. $8x^2 - x - 11$
3a. $7ab$
b. 0
4a. $-2a$
b. $-x^2 - x + 4$
c. $-2x^2 - 9x - 9$
5a. a^3
b. y^{15}
c. $x^2 y^4 z^6$
d. $-125 a^3 x^6$
6a. $-3x^3 y^2$
b. $2x^3 - 6x^2 + 10x$
c. $x^2 - 13x + 42$
d. $15x^3 + 16x^2 - 8x - 8$
7a. x^3
b. y^2
c. $1/x^3$
d. z/x
8a. $2xy$
b. $\frac{4}{7y^2} + \frac{6}{7y} - \frac{2x}{7y}$
c. $2x - 5$

CHAPTER 6 TEST F

1a. $3x$
b. $-4z$
c. $11a^3$
2a. $7x^2 y + 3x + y^2$
b. $9x^2 + 4y^2$
c. $7x + 6y$
3a. $7x^2$
b. $4ab$
4a. $-7x^2 - y$
b. $4 + x$
c. $-2x^2 - 9x - 9$
5a. y^6
b. y^{15}
c. $x^{12} y^4$
d. x^8
6a. $2x^3 y^2$
b. $-3x^3 - 12x^2 + 3x$
c. $4x^2 + 4x + 1$
d. $2x^3 - 9x^2 - 6x + 1$
7a. x^6
b. z
c. $\frac{1}{y^3 x^2}$
d. $x^5 y / z^3$
8a. $2y^2 / 3x$
b. $\frac{x}{5} - \frac{1}{3}$
c. $2x + 3$

CHAPTER 7 TEST A

1a. $4(x^2 + y^2)$
b. $3x^2(1 - 7x)$
c. $(y-1)(y+3)$
2a. $a^2 - 2a - 15$
b. $2x^2 - 11x + 5$
c. $2z^2 + z - 6$
d. $4z^2 - 4z + 1$
3a. $(x+5)(x+1)$
b. $(x-7)(x-5)$
c. $(x+8)(x-6)$
d. $(x-15)(x+2)$
4a. $(2x+1)(x+5)$
b. $(2x-5)(x-2)$
c. $(4x+3)(x+5)$
d. $(3x-2)(2x-3)$
5a. $(x+5)(x-5)$
b. $(2x+1)(2x-1)$
6a. $3(x^2 + 3x + 4)$
b. $3(x+3)(x-3)$
c. $2x(x-3)(x+2)$
7a. $x = 0, -2$
b. $x = 3, -3$
c. $x = 3, 1$
d. $x = -5/2, -2$

CHAPTER 7 TEST B

1a. $3(a^2 + 3)$
b. $-3y(1 + 2y)$
c. $(x+1)(x+2)$
2a. $x^2 + 2x + 1$
b. $2x^2 - 7x + 5$
c. $5y^2 + 7y - 6$
d. $4x^2 - 7x + 9$
3a. $(x+4)(x+1)$
b. $(x-8)(x+1)$
c. $(x+3)(x-2)$
d. $(x-9)(x-1)$
4a. $(2x+3)(x+2)$
b. $(3x-2)(x-1)$
c. $2(x-1)(2x+3)$
d. $(4x-3)(5x-1)$
5a. $(z+10)(z-10)$
b. $(3x+7)(3x-7)$
6a. $7(x-3)(x+1)$
b. $(x^2+1)(x+1)(x-1)$
c. $3a(x-2)(x-1)$
7a. $x = 0, 5$
b. $x = 5, -5$
c. $x = 2, -7$
d. $x = -7/3, -2$

CHAPTER 7 TEST C	CHAPTER 7 TEST D	CHAPTER 7 TEST E	CHAPTER 7 TEST F
1a. $6(2c^2+1)$	1a. $9(2x^2-3x+1)$	1a. $-2(2x^2+2x-5)$	1a. $-7(x^2-xy+y^2)$
b. $2x(x^2-5x-2)$	b. $2ab(1+3c^2)$	b. $x(x^2-x-1)$	b. $2y(3y-2)$
c. $(x-1)(2x-3)$	c. $(y-1)(y-3)$	c. $(z-5)(z+1)$	c. $(2x+1)(x-1)$
2a. $y^2-3y-28$	2a. z^2+z-30	2a. x^2-4x+3	2a. y^2+8y-9
b. $x^2-17x+35$	b. $6y^2-y-12$	b. $5x^2-11x-12$	b. $2x^2+9x+4$
c. $x^2+11x-4$	c. $4x^2-4x+1$	c. $4x^2-9$	c. $9x^2-6x+1$
d. $9x^2+6x+1$	d. $9x^2+24x+16$	d. x^2-2x+1	d. x^2+2x+1
3a. $(x+5)(x+2)$	3a. $(x+7)(x+4)$	3a. $(x+8)(x+5)$	3a. $(x+5)(x+3)$
b. $(y+4)(y-3)$	b. $(x-5)(x+4)$	b. $(x-1)(x-2)$	b. $(x+9)(x-8)$
c. $(x-12)(x-12)$	c. $(z+3)(z-2)$	c. $(x-3)(x+1)$	c. $(x-3)(x+2)$
d. $(x+3)(x-8)$	d. $(x-10)(x-2)$	d. $(x+4)(x-3)$	d. $(x-7)(x-5)$
4a. $(3x-5)(x+2)$	4a. $(2x-3)(x-3)$	4a. $(7x-8)(x+1)$	4a. $(3x-4)(x+3)$
b. $(5x-7)(x+1)$	b. $(3x-1)(x+7)$	b. $(13x+6)(x-1)$	b. $(3x+4)(x+2)$
c. $(8x+3)(x-2)$	c. $(4x+6)(x-3)$	c. $(2y+1)(3y-5)$	c. $(2x-1)(5x-2)$
d. $(2x+5)(3x+7)$	d. $(2x-1)(3x+5)$	d. $(5x-2)(3x+4)$	d. $(2x-3)(3x+4)$
5a. $(x+8)(x-8)$	5a. $(5x+y)(5x-y)$	5a. $(3x+2y)(3x-2y)$	5a. $(2x+5)(2x-5)$
b. $(5x+7)(5x-7)$	b. $(3x+1)(3x-1)$	b. $(5+x)(5-x)$	b. $(4x+10y)(4x-10y)$
6a. $5(6x-1)(x+1)$	6a. $4(2x+1)(x-3)$	6a. $2(x-5)(x-2)$	6a. $2(y-5)(y+3)$
b. $4x(y+3)(y-1)$	b. $x(x+3)(x-3)$	b. $x(x+1)(x-1)$	b. $3x(9x^2+2x+1)$
c. $2(3x+2)(3x-2)$	c. $2x(x+5)(x-2)$	c. $2y(x-6)(x-6)$	c. $x^2(x^2+1)(x+1)(x-1)$
7a. $x=0, -7/2$	7a. $x=0, 4$	7a. $x=0, 5$	7a. $x=0, -5$
b. $x=3, -3$	b. $x=10, -10$	b. $x=3, -3$	b. $x=5, -5$
c. $x=5, 1$	c. $x=8, -3$	c. $x=-2, -1$	c. $x=1, -6$
d. $x=3/2, 7$	d. $x=-1/3, -3$	d. $x=1/2$	d. $x=-4/3, 5/2$

CHAPTER 8 TEST A

1a. $3x^2$
b. $(x+y)/(x-y)$
c. $(x-7)/2$
d. $x/(x+3)$
e. $-1/(x+3)$
2a. $y/8$
b. $8/xy$
c. $x^2(x-y)/2(x+y)$
d. $(x-7)/(y+5)$
3a. x/y
b. $9/14$
c. $(x+y)/xy$
d. 1
4a. $5/(x-1)$
b. 1
c. $(2x-4)/7$
5a. $(4r+15)/5r$
b. $(2x-7)/12x$
c. $1/6(x-y)$
d. $4a/3$
6a. $x^2/(x+1)$
b. $4(x+1)/9x$
7a. $x=9$
b. $x=2$
8. 45 MINUTES

CHAPTER 8 TEST B

1a. x
b. $1/2$
c. $(x+y)/(x-y)$
d. $(x+4)/(x-3)$
e. $-1/(x+1)$
2a. $3cd/4ab$
b. $33x$
c. $24xy/15(x+y)$
d. $-(3-a)/45$
3a. $5x^3y^2/2$
b. $x^2/3$
c. $3(x+1)/4$
d. $(x-1)/(x+1)$
4a. $7x/(x+1)$
b. $-2/x$
c. $(3x+8)/5$
5a. $13/12x$
b. $(2x+y)/xy$
c. $4(x-3)/x(x-2)$
d. $(3x+5)/(x+1)$
6a. $(2x+3)/x^2$
b. $1/5$
7a. $x=6$
b. $x=4$
8. 1 HR., 36 MIN.

CHAPTER 8 TEST C

1a. $-4/3x^2y^2$
b. $1/7$
c. $(x+4)/4$
d. $(x-1)/(x+2)$
e. $-(x+5)$
2a. $2y^2$
b. $ab(x+y)$
c. $1/x$
d. 1
3a. $4ax/5$
b. $ab/4c$
c. $12(x-1)/(x+1)^2$
d. $2(4x-y)/(x+3)$
4a. $(x+3)/(x-1)$
b. $2x$
c. $1/(x+1)$
5a. $13/5x$
b. $-3(x-1)/4$
c. $\dfrac{-2x+3y}{(x+y)(x-y)}$
d. $\dfrac{(y+2)(y-2)}{y}$
6a. $(x+4)/x$
b. $3/10$
7a. $x=-10$
b. $x=-10$
8. $4/5$ HR. $\left(\dfrac{48}{MIN}\right)$

CHAPTER 8 TEST D

1a. $8/z$
b. $1/5$
c. $4/(x-y)$
d. $3/(x-2)$
e. $-(x-3)/(x+3)$
2a. $5b/7a^2$
b. $2r$
c. $3/x$
d. $3/4$
3a. $4x/5$
b. $24y^2/5$
c. $2(x+1)/(x-3)$
d. $x(x+8)/16$
4a. $4x/(x+2)$
b. $x+4$
c. $(2x-1)/y$
5a. $(3-2a)/a^2$
b. $(5a-1)/4$
c. $\dfrac{5x-1}{2(3x+1)(3x-1)}$
d. $x/4$
6a. y/x
b. $2(2xy-1)/(xy-1)$
7a. $x=11$
b. $x=1/2$
8. 2.1 MINUTES

CHAPTER 8 TEST E

1a. $y/6$
b. $3(x+y)/4(x+z)$
c. $(x-5)/3$
d. $(3x+1)/(3x-1)$
e. $-1/6$
2a. $8y$
b. $2/y$
c. $(x-4)/10x$
d. $(x-3)/5x$
3a. $xy/12$
b. $15a^2$
c. $2(x-3)$
d. $\dfrac{(x+3)(x-3)}{8(x-2)}$
4a. x
b. $1/3$
c. $1/(x-1)$
5a. $(10y+x)/8xy$
b. $(5x-1)/6x$
c. $\dfrac{x(8x+19)}{(x+2)(x+3)}$
d. $2(5y-1)/(2y-1)$
6a. $2(x-3)/x(3x+1)$
b. $(x-y)/(x+y)$
7a. $x=8$
b. $x=2$
8. $1\frac{1}{2}$ HR.

CHAPTER 8 TEST F

1a. $-3/4xy$
b. $(x-y)/x^2$
c. $2(x-1)$
d. $(x-4)/(x-3)$
e. $-(a+b)/3$
2a. $5xy/6$
b. 4
c. $(1-2x)/9$
d. $-(x+1)/x$
3a. y^4/x^2
b. $x/30y^2$
c. $(x+1)/2$
d. $\dfrac{6}{(x+2)(x-2)}$
4a. 1
b. $-1/2$
c. $(a+2b)/3x$
5a. $\dfrac{3y-10x}{18xy}$
b. $-(x-2)/x$
c. $1/2(x-5)$
d. $7x/5$
6a. $\dfrac{y(x+3y)}{3(x-3y)}$
b. $4/(3x-1)$
7a. $x=-24$
b. $x=-8$
8. 1.2 HR.

CHAPTER 9 TEST A

1a. x
b. $5y^2 x$
2a. $5\sqrt{2}$
b. $12\sqrt{2}$
c. $xy\sqrt{y}$
d. $3x^2 y\sqrt{2x}$
3a. $3/4$
b. $2\sqrt{2}/5$
c. $2\sqrt{5}$
4a. $8\sqrt{3}$
b. $4\sqrt{2}+3\sqrt{5}$
c. $\sqrt{2}$
5a. 4
b. $20\sqrt{3}$
c. 12
d. $\sqrt{6}+\sqrt{10}$
6a. $\sqrt{5}$
b. $\sqrt{2}/2$
c. $\sqrt{3}/3$
d. $\sqrt{6}$
7a. $x=36$
b. NO SOLUTION

CHAPTER 9 TEST B

1a. y^3
b. $12x^2 y$
2a. $2\sqrt{5}$
b. $-6\sqrt{6}$
c. $x^2 y^2\sqrt{xy}$
d. $4x\sqrt{3x}$
3a. $1/3$
b. $3\sqrt{2}/4$
c. $2\sqrt{6}$
4a. $3\sqrt{5}$
b. $5\sqrt{2}-6\sqrt{3}$
c. $2\sqrt{2}$
5a. 5
b. $10\sqrt{10}$
c. $28x^3$
d. $50\sqrt{2}-25\sqrt{6}$
6a. $2\sqrt{6}$
b. $\sqrt{3}/3$
c. $7/4$
d. $\sqrt{15}/2$
7a. $x=9$
b. $x=27$

CHAPTER 9 TEST C	CHAPTER 9 TEST D	CHAPTER 9 TEST E	CHAPTER 9 TEST F
1a. $2x^3$	1a. $4y^2$	1a. $7z^5$	1a. $8x^3$
b. $5x^3y^4$	b. x^3yz^2	b. $2x^2z^2$	b. $3xy^2z^3$
2a. $3\sqrt{10}$	2a. $2\sqrt{11}$	2a. $5\sqrt{5}$	2a. $4\sqrt{2}$
b. $2\sqrt{3}$	b. $\sqrt{5}$	b. $8\sqrt{17}$	b. $-2\sqrt{3}$
c. $x^2y^2\sqrt{y}$	c. $xy^2\sqrt{x}$	c. $x^3\sqrt{x}$	c. $xy^2\sqrt{z}$
d. $-6x\sqrt{5y}$	d. $6xy\sqrt{2xy}$	d. $6xy^2\sqrt{6xy}$	d. $6xy^2\sqrt{y}$
3a. 4	3a. $3/7$	3a. 4	3a. $2/3$
b. $3\sqrt{6}/7$	b. $2\sqrt{5}/5$	b. $\sqrt{2}/2$	b. $2\sqrt{2}/3$
c. $3\sqrt{7}$	c. $10\sqrt{3}/3$	c. $2\sqrt{22}/3$	c. $12\sqrt{2}$
4a. $-3\sqrt{7}$	4a. $5\sqrt{3}$	4a. $12\sqrt{2}$	4a. $7\sqrt{3}$
b. $3\sqrt{6}-\sqrt{7}$	b. $2\sqrt{2}+\sqrt{3}$	b. $\sqrt{5}+3\sqrt{6}$	b. $10\sqrt{7}$
c. $7\sqrt{3}$	c. $10\sqrt{3}$	c. $-11\sqrt{2}$	c. 0
5a. 6	5a. 7	5a. 10	5a. x
b. $21x$	b. $4x^2\sqrt{3}$	b. $-10x^2\sqrt{x}$	b. $9a$
c. $4x$	c. $2x+1$	c. $75x$	c. $4x^3$
d. $\sqrt{6}-\sqrt{21}$	d. $12\sqrt{3}+6\sqrt{2}$	d. $3\sqrt{2}$	d. $-8\sqrt{15}$
6a. $\sqrt{2}/3$	6a. $4\sqrt{2}$	6a. $2\sqrt{2}/3$	6a. $3\sqrt{5}$
b. $\sqrt{10}/5$	b. $\sqrt{14}/7$	b. $3\sqrt{5}/5$	b. $\sqrt{30}/3$
c. $\sqrt{3}$	c. $\sqrt{2}/2$	c. $5\sqrt{2}$	c. $\sqrt{30}/5$
d. $8\sqrt{6}/3$	d. $10\sqrt{3}/3$	d. $\sqrt{35}$	d. $\sqrt{6}$
7a. NO SOLUTION	7a. $x=6$	7a. $x=54$	7a. NO SOLUTION
b. $x=12$	b. $x=10$	b. $x=12$	b. $y=3$

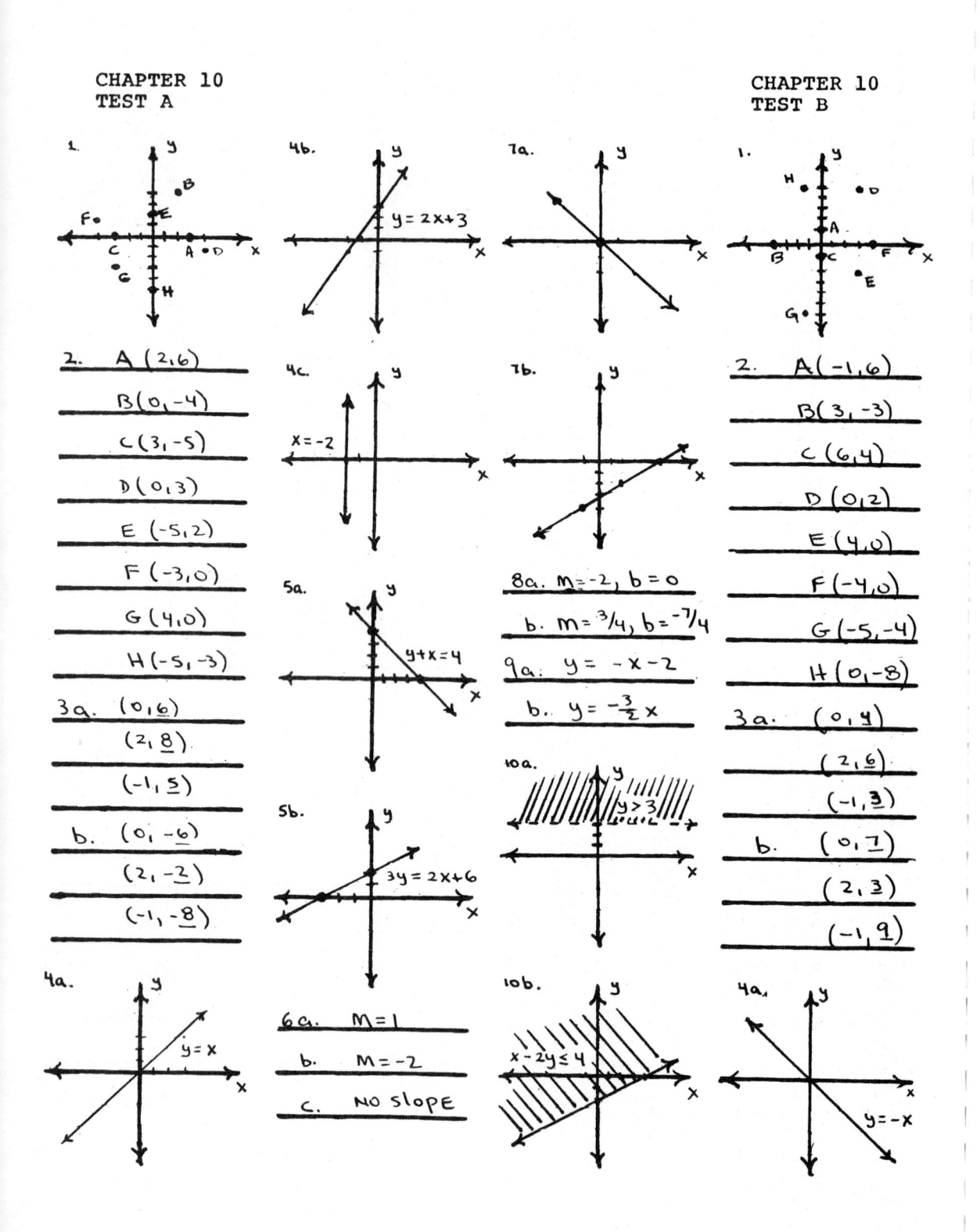

CHAPTER 10 TEST A

1. [graph with points A, B, C, D, E, F, G, H]

2. A (2,6)
   B (0,-4)
   C (3,-5)
   D (0,3)
   E (-5,2)
   F (-3,0)
   G (4,0)
   H (-5,-3)

3a. (0,6)
    (2,8)
    (-1,5)

 b. (0,-6)
    (2,-2)
    (-1,-8)

4a. [graph of $y = x$]

4b. [graph of $y = 2x + 3$]

4c. [graph of $x = -2$]

5a. [graph of $y + x = 4$]

5b. [graph of $3y = 2x + 6$]

6a. M = 1
 b. M = -2
 c. NO SLOPE

7a. [graph]

7b. [graph]

8a. $m = -2$, $b = 0$
 b. $m = 3/4$, $b = -7/4$

9a. $y = -x - 2$
 b. $y = -\frac{3}{2}x$

10a. [graph of $y > 3$]

10b. [graph of $x - 2y \leq 4$]

CHAPTER 10 TEST B

1. [graph with points A, B, C, D, E, F, G, H]

2. A (-1,6)
   B (3,-3)
   C (6,4)
   D (0,2)
   E (4,0)
   F (-4,0)
   G (-5,-4)
   H (0,-8)

3a. (0,4)
    (2,6)
    (-1,3)

 b. (0,7)
    (2,3)
    (-1,9)

4a. [graph of $y = -x$]

CHAPTER 10
TEST C

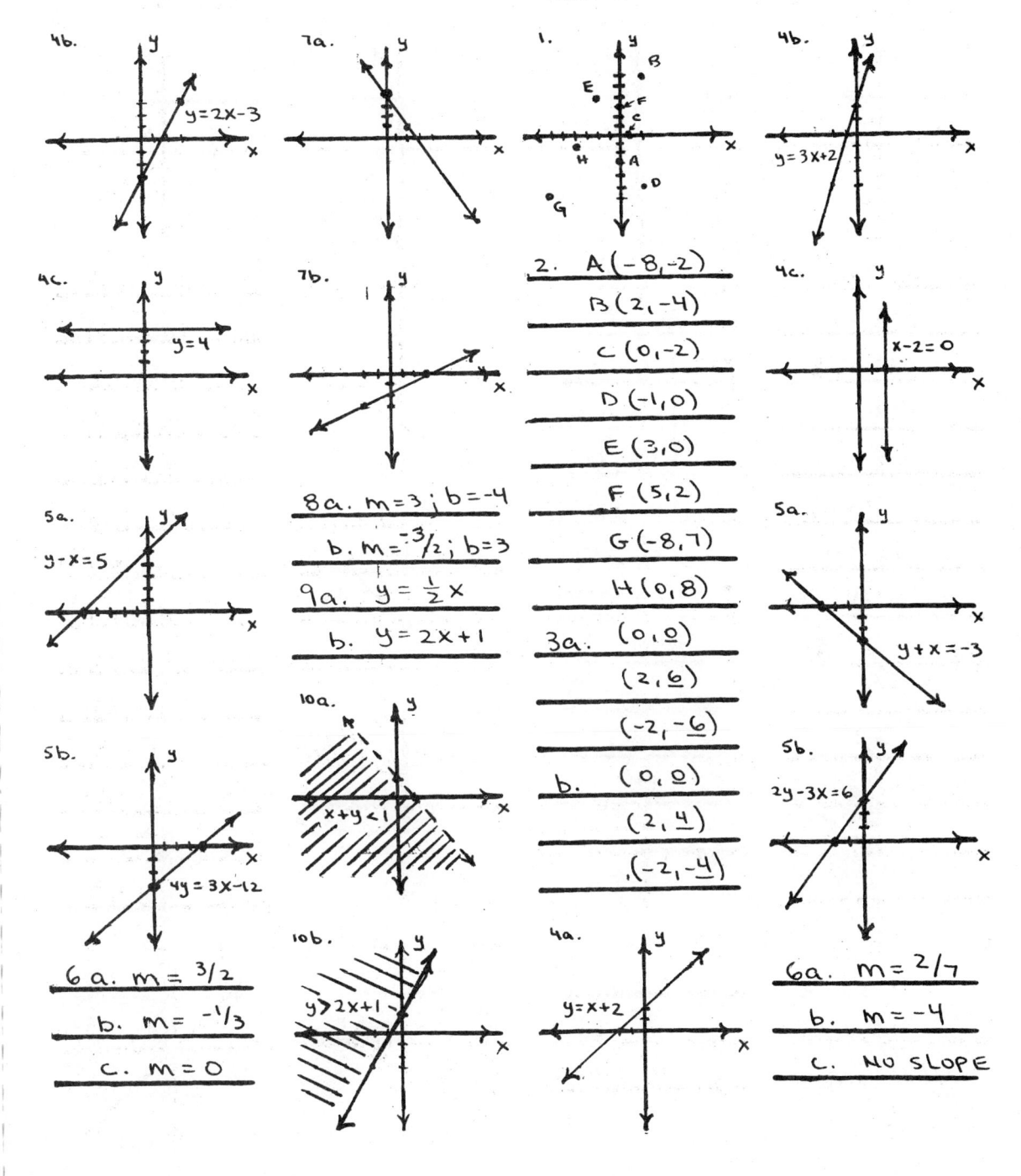

2. A $(-8,-2)$
   B $(2,-4)$
   C $(0,-2)$
   D $(-1,0)$
   E $(3,0)$
   F $(5,2)$
   G $(-8,7)$
   H $(0,8)$

3a. $(0,\underline{0})$
    $(2,\underline{6})$
    $(-2,\underline{-6})$

b. $(0,\underline{0})$
   $(2,\underline{4})$
   $(-2,\underline{-4})$

6a. $m = 3/2$
b. $m = -1/3$
c. $m = 0$

6a. $m = 2/7$
b. $m = -4$
c. NO SLOPE

8a. $m = 3$; $b = -4$
b. $m = -3/2$; $b = 3$

9a. $y = \frac{1}{2}x$
b. $y = 2x+1$

CHAPTER 10
TEST D

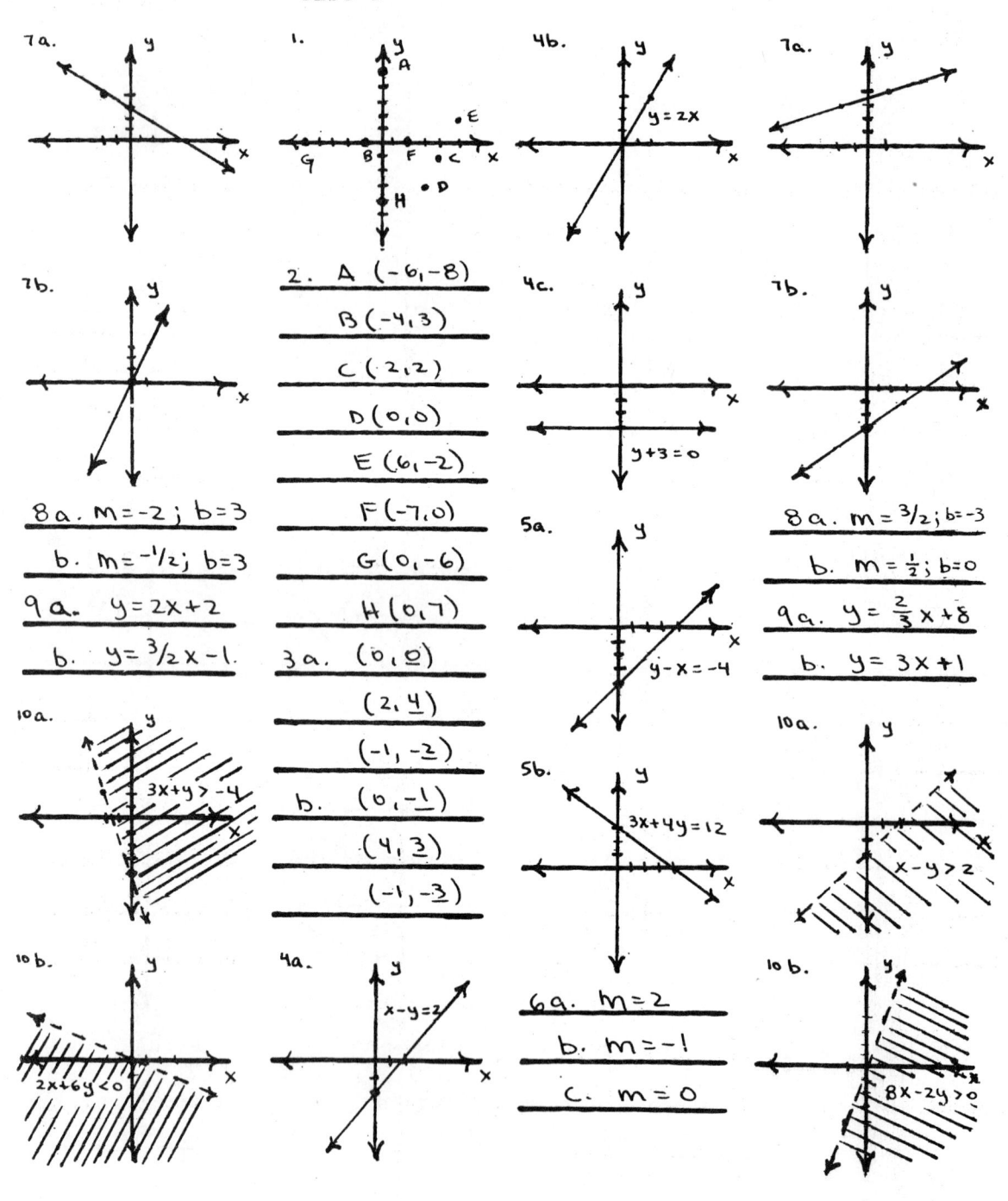

7a. [graph]

1. [graph with points A, B, C, D, E, F, G, H]

4b. [graph of y=2x]

7a. [graph]

7b. [graph]

2. A (−6, −8)
B (−4, 3)
C (2, 2)
D (0, 0)
E (6, −2)
F (−7, 0)
G (0, −6)
H (0, 7)

4c. [graph of y+3=0]

7b. [graph]

8a. m = −2; b = 3
b. m = −1/2; b = 3
9a. y = 2x + 2
b. y = 3/2 x − 1

3a. (0, <u>0</u>)
(2, <u>4</u>)
(−1, <u>−2</u>)
b. (0, <u>−1</u>)
(4, <u>3</u>)
(−1, <u>−3</u>)

5a. [graph of y − x = −4]

8a. m = 3/2; b = −3
b. m = 1/2; b = 0
9a. y = 2/3 x + 8
b. y = 3x + 1

10a. [graph of 3x+y > −4]

5b. [graph of 3x+4y=12]

10a. [graph of x−y > 2]

10b. [graph of 2x+6y < 0]

4a. [graph of x−y=2]

6a. m = 2
b. m = −1
c. m = 0

10b. [graph of 8x−2y > 0]

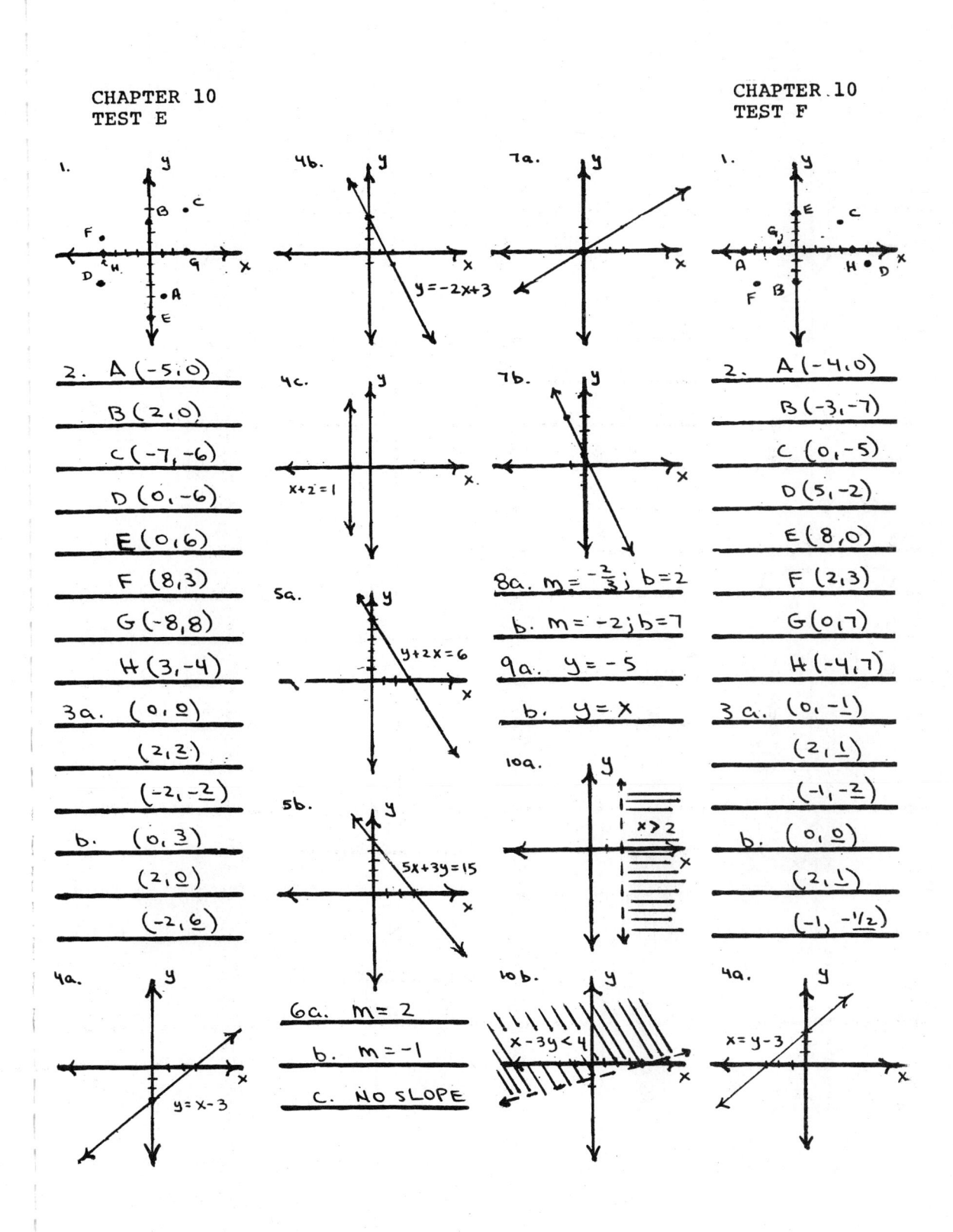

CHAPTER 10 TEST E

1. [graph with points B, C, F, D, H, G, A, E]

2. A(-5,0)
   B(2,0)
   C(-7,-6)
   D(0,-6)
   E(0,6)
   F(8,3)
   G(-8,8)
   H(3,-4)

3a. (0, 0)
    (2, 3)
    (-2, -2)
  b. (0, 3)
    (2, 0)
    (-2, 6)

4a. [graph of y = x - 3]

4b. [graph of y = -2x + 3]

4c. [graph of x + 2 = 1]

5a. [graph of y + 2x = 6]

5b. [graph of 5x + 3y = 15]

6a. m = 2
  b. m = -1
  c. NO SLOPE

7a. [graph]

7b. [graph]

8a. $m = -\frac{2}{3}$; b = 2
  b. m = -2; b = 7

9a. y = -5
  b. y = x

10a. [graph, x > 2]

10b. [graph, x - 3y < 4]

CHAPTER 10 TEST F

1. [graph with points E, C, G, A, H, D, F, B]

2. A(-4,0)
   B(-3,-7)
   C(0,-5)
   D(5,-2)
   E(8,0)
   F(2,3)
   G(0,7)
   H(-4,7)

3a. (0, -1)
    (2, 1)
    (-1, -2)
  b. (0, 0)
    (2, 1)
    (-1, -1/2)

4a. [graph of x = y - 3]

CHAPTER 11 TEST A

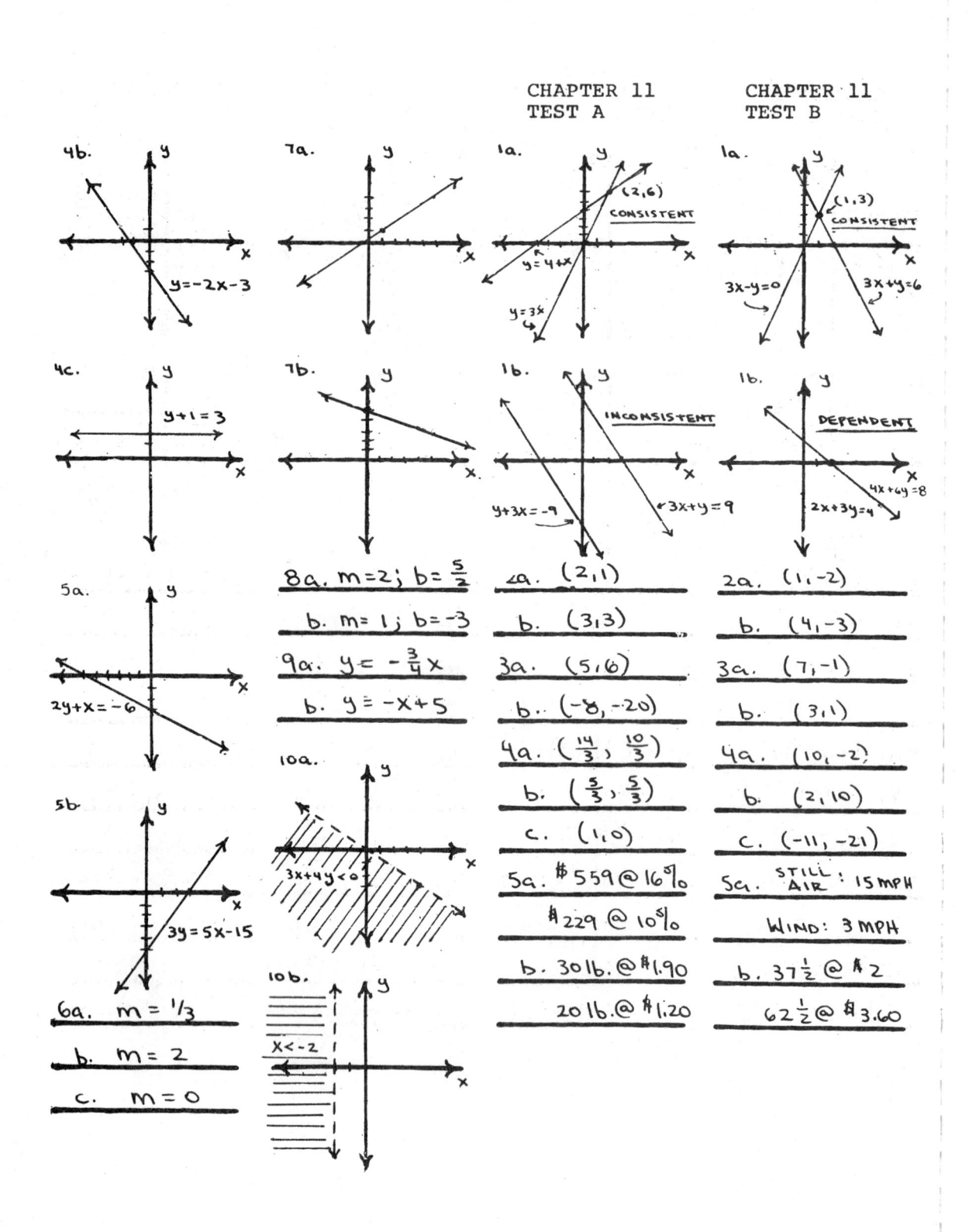

1a. (2,6) CONSISTENT; lines $y=4+x$ and $y=3x$

1b. INCONSISTENT; lines $y+3x=-9$ and $3x+y=9$

2a. (2,1)
b. (3,3)

3a. (5,6)
b. (-8,-20)

4a. $(\frac{14}{3}, \frac{10}{3})$
b. $(\frac{5}{3}, \frac{5}{3})$
c. (1,0)

5a. $559 @ 16%
   $229 @ 10%
b. 30 lb. @ $1.90
   20 lb. @ $1.20

CHAPTER 11 TEST B

1a. (1,3) CONSISTENT; lines $3x-y=0$ and $3x+y=6$

1b. DEPENDENT; lines $4x+6y=8$ and $2x+3y=4$

2a. (1,-2)
b. (4,-3)

3a. (7,-1)
b. (3,1)

4a. (10,-2)
b. (2,10)
c. (-11,-21)

5a. STILL AIR: 15 MPH
    WIND: 3 MPH
b. $37\frac{1}{2}$ @ $2
   $62\frac{1}{2}$ @ $3.60

4b. graph of $y=-2x-3$

4c. graph of $y+1=3$

5a. graph of $2y+x=-6$

5b. graph of $3y=5x-15$

6a. $m=\frac{1}{3}$
b. $m=2$
c. $m=0$

7a. graph (line through origin, positive slope)

7b. graph (line with negative slope)

8a. $m=2;\ b=\frac{5}{2}$
b. $m=1;\ b=-3$

9a. $y=-\frac{3}{4}x$
b. $y=-x+5$

10a. graph of $3x+4y<0$

10b. graph of $x<-2$

CHAPTER 11 TEST C

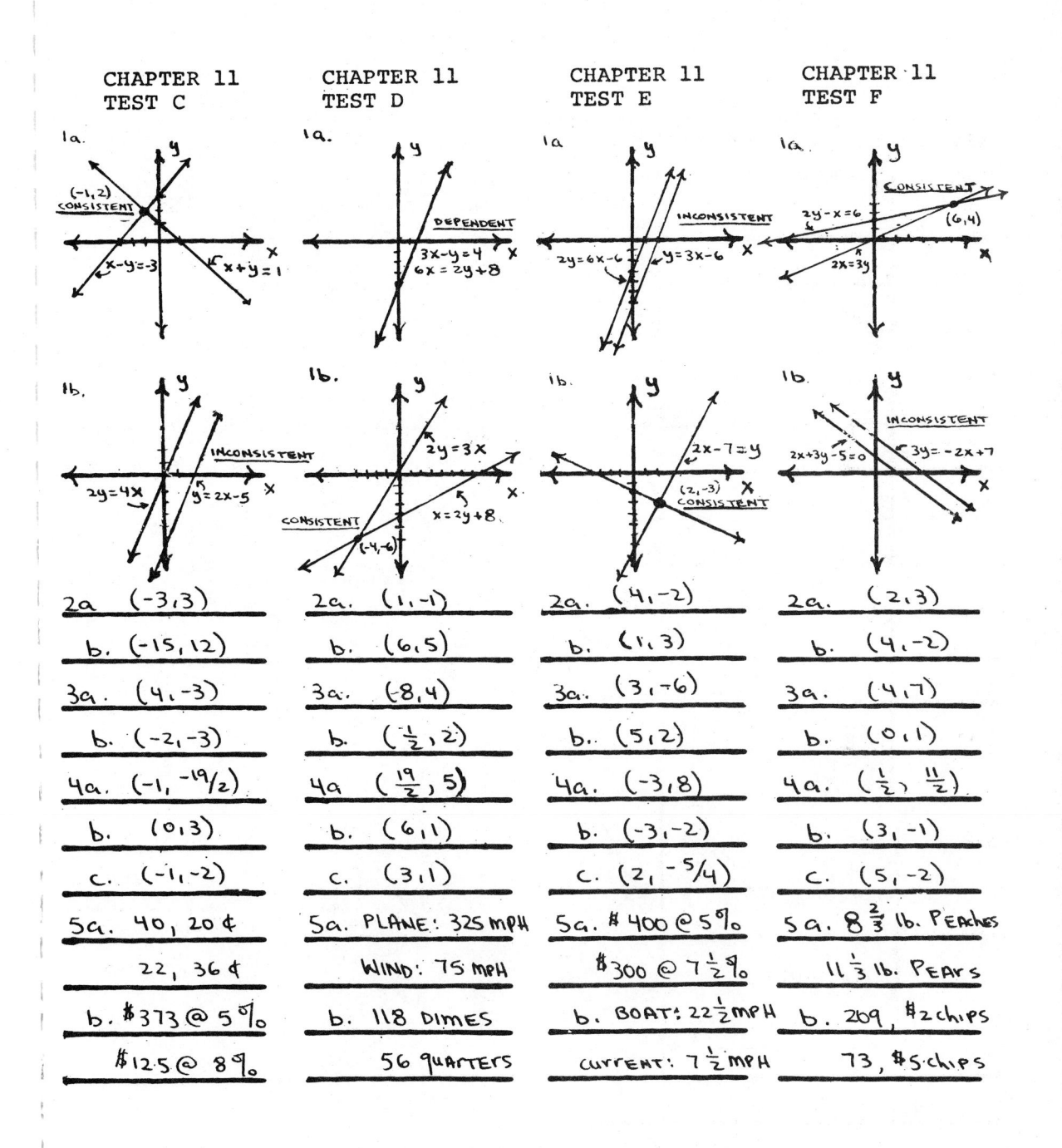

1a. (graph) $(-1,2)$ CONSISTENT; $x-y=-3$, $x+y=1$

1b. (graph) INCONSISTENT; $2y=4x$, $y=2x-5$

2a. $(-3,3)$
b. $(-15,12)$
3a. $(4,-3)$
b. $(-2,-3)$
4a. $(-1, -19/2)$
b. $(0,3)$
c. $(-1,-2)$
5a. 40, 20¢
   22, 36¢
b. $373 @ 5%
   $125 @ 8%

CHAPTER 11 TEST D

1a. (graph) DEPENDENT; $3x-y=4$, $6x=2y+8$

1b. (graph) CONSISTENT; $2y=3x$, $x=2y+8$, $(-4,-6)$

2a. $(1,-1)$
b. $(6,5)$
3a. $(-8,4)$
b. $(\frac{1}{2}, 2)$
4a. $(\frac{19}{2}, 5)$
b. $(6,1)$
c. $(3,1)$
5a. PLANE: 325 MPH
   WIND: 75 MPH
b. 118 DIMES
   56 QUARTERS

CHAPTER 11 TEST E

1a. (graph) INCONSISTENT; $2y=6x-6$, $y=3x-6$

1b. (graph) CONSISTENT; $2x-7=y$, $(2,-3)$

2a. $(4,-2)$
b. $(1,3)$
3a. $(3,-6)$
b. $(5,2)$
4a. $(-3,8)$
b. $(-3,-2)$
c. $(2, -5/4)$
5a. $400 @ 5%
   $300 @ 7\frac{1}{2}%
b. BOAT: $22\frac{1}{2}$ MPH
   CURRENT: $7\frac{1}{2}$ MPH

CHAPTER 11 TEST F

1a. (graph) CONSISTENT; $2y-x=6$, $2x=3y$, $(6,4)$

1b. (graph) INCONSISTENT; $2x+3y-5=0$, $3y=-2x+7$

2a. $(2,3)$
b. $(4,-2)$
3a. $(4,7)$
b. $(0,1)$
4a. $(\frac{1}{2}, \frac{11}{2})$
b. $(3,-1)$
c. $(5,-2)$
5a. $8\frac{2}{3}$ lb. PEACHES
   $11\frac{1}{3}$ lb. PEARS
b. 209, $2 chips
   73, $5 chips

CHAPTER 12 TEST A

1a. $2x^2 - 8x - 11 = 0$
b. $2x^2 - 6x + 3 = 0$
2a. $x = 0, -5$
b. $x = 4, -4$
c. $x = -7, 3$
d. $x = \frac{3}{2}, 2$
3a. $x = \pm\sqrt{3}$
b. $x = -1, -5$
4a. $x = 1, -5$
b. $x = -1, 4$
c. $x = 3, -5$
5a. $x = -3 \pm \sqrt{6}$
b. $x = 2, \frac{2}{3}$

6a. Graph of $y = x^2 - 6x + 5$, vertex $(3, -4)$
6b. Graph of $y = x^2 - 4$, vertex $(0, -4)$

CHAPTER 12 TEST B

1a. $4x^2 + 4x - 5 = 0$
b. $x^2 - 12x + 4 = 0$
2a. $x = 0, 2$
b. $x = 3, -3$
c. $x = 4, -2$
d. $x = -\frac{1}{2}$
3a. $x = \pm 2\sqrt{2}$
b. $x = 4, -2$
4a. $x = -2, -4$
b. $x = 4, -1$
c. $x = 1 \pm \frac{\sqrt{6}}{3}$
5a. $x = 5 \pm \sqrt{30}$
b. $x = \frac{-6 \pm \sqrt{57}}{3}$

6a. Graph of $y = x^2 - 4x$, vertex $(2, -4)$
6b. Graph of $y = 2x^2 - 2x - 12$, vertex $(1/2, -25/2)$

CHAPTER 12 TEST C

1a. $y^2 + y - 30 = 0$
b. $x^2 - 6x - 27 = 0$
2a. $x = 0, 5$
b. $x = 5, -5$
c. $x = 6, 2$
d. $x = \frac{3}{4}, -6$
3a. $x = \pm 2\sqrt{2}$
b. $x = \frac{1 \pm 2\sqrt{2}}{2}$
4a. $x = 7, -2$
b. $x = 3 \pm 2\sqrt{2}$
c. $x = \frac{-3 \pm \sqrt{11}}{2}$
5a. $x = \frac{-9 \pm \sqrt{97}}{2}$
b. $x = \frac{1}{2}, -3$

6a. Graph of $y = x^2 - 9$, vertex $(0, -9)$
6b. Graph of $y = 3x^2 - 2x - 8$, vertex $(1/3, -25/3)$

CHAPTER 12 TEST D

1a. $8x^2 - 16x - 10 = 0$
b. $2x^2 + 6x - 1 = 0$
2a. $x = 0, -5$
b. $y = 5, -5$
c. $x = -9, -1$
d. $x = \frac{7}{3}, 1$
3a. $x = \pm 2\sqrt{6}$
b. $x = 3, -5$
4a. $x = \frac{9 \pm 3\sqrt{5}}{2}$
b. $x = -3 \pm \sqrt{10}$
c. $x = 2 \pm \frac{\sqrt{10}}{2}$
5a. $x = 10 \pm 3\sqrt{10}$
b. $x = \frac{-5 \pm \sqrt{13}}{2}$

6a. Graph of $y = x^2 - 6x - 16$, vertex $(3, -25)$
6b. Graph of $y = x^2 - 4x + 3$, vertex $(2, -1)$

CHAPTER 12 TEST E

1a. $x^2 - 2x - 1 = 0$
b. $2x^2 - 9 = 0$
2a. $x = 0, -3$
b. $x = 3, -3$
c. $x = 5, -1$
d. $x = \frac{1}{2}, -\frac{9}{2}$
3a. $x = \pm 3\sqrt{3}$
b. $y = -3, -7$
4a. $x = 5 \pm \sqrt{30}$
b. $x = 5, -2$
c. $x = \frac{9 \pm \sqrt{165}}{6}$
5a. $x = \frac{-1 \pm \sqrt{21}}{2}$
b. $x = \frac{-1 \pm \sqrt{33}}{2}$

CHAPTER 12 TEST F

1a. $9z^2 - 90z + 176 = 0$
b. $x^2 - 25 = 0$
2a. $x = 0, 1$
b. $x = 2, -2$
c. $x = 10, -2$
d. $x = \frac{1}{2}, 2$
3a. $x = \pm 4\sqrt{2}$
b. $x = 3 \pm 2\sqrt{3}$
4a. $x = -2 \pm \sqrt{7}$
b. $x = 9, -2$
c. $x = \frac{15 \pm \sqrt{249}}{6}$
5a. $x = \frac{-7 \pm \sqrt{51}}{2}$
b. $x = \frac{1 \pm \sqrt{19}}{3}$

6a. $y = x^2 - 2x - 3$; vertex $(1, -4)$

6b. $y = x^2 + x - 12$; vertex $(-\frac{1}{2}, -\frac{49}{4})$

6a. $y = x^2 - 6x - 7$; vertex $(3, -16)$

6b. $y = x^2 + 6x + 5$; vertex $(-3, -4)$

FINAL EXAMINATIONS
ANSWERS

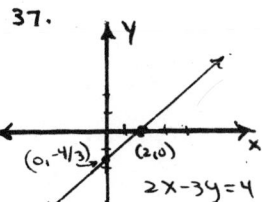

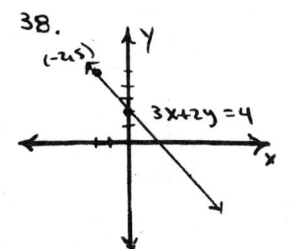

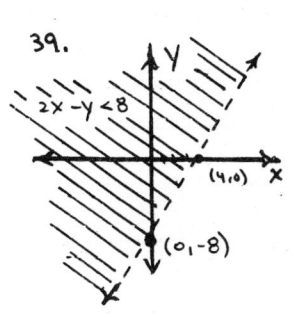

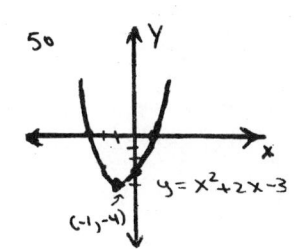

FINAL EXAMINATION
FORM A

1. 19
2. −9
3. 0
4. 44
5. $-30/7$
6. $x = 10$
7. $x = -18/7$
8. $x = 4$
9. $x = 5/7$
10. $h = \dfrac{V}{\pi r^3}$
11. $x \geq -1$ [number line: closed at −1, arrow right]
12. $x < 4$ [number line: open at 4, arrow left]
13. (a) 14 lb. CASHEWS
       36 lb. WALNUTS

   (b) \$6000 @ 8%
       \$4000 @ 6%

   (c) Ⓐ 70 MPH
       Ⓑ 55 MPH

14. $4x^3 - 8x^2 + 2x - 5$
15. $6x^2 + 2x - 20$
16. $3x^3 + 16x^2 - 4x + 14$
17. $\dfrac{6}{x^2 y} - 3x^2 y^2 + \dfrac{8z}{x^3}$
18. $x^2 + 6x - 4$
19. $5ab(1 - 2b + 3a)$
20. $(x+6)(x-4)$
21. $(2x+3)(x-5)$
22. $x(2x+3)(3x-4)$
23. $(2x+5y)(2x-5y)$
24. $\dfrac{2(x-7)}{5(x-3)}$
25. $\dfrac{-(x-1)}{x}$
26. $\dfrac{x+13}{2x}$
27. $\dfrac{-x^2 + 8x - 10}{(x+3)(x-2)(x+4)}$
28. $1/2$
29. $x = 8$
30. $x = -3/95$
31. $x = 2$
32. $6\sqrt{2}$
33. $-2\sqrt{3}$
34. $-\sqrt{2}$
35. $\dfrac{2\sqrt{3}}{5}$
36. $\dfrac{3\sqrt{30}}{20}$

37. [graph of line $2x - 3y = 4$ through $(0, -4/3)$ and $(2, 0)$]

38. [graph of line $3x + 2y = 4$ through $(-2, 5)$]

39. [graph of region $2x - y < 8$, with intercepts $(4, 0)$ and $(0, -8)$, shaded]

40. $y = 5/2\, x$
41. $y = 3x + 1$
42. $(2, -2/3)$
43. $(12, 3)$
44. $(-41, 25)$
45. $x = 0, 4$
46. $x = 3, -3$
47. $x = 3, -1$
48. $-1 \pm \sqrt{2}$
49. $x = 1/3, -1$

50. [graph of parabola $y = x^2 + 2x - 3$ with vertex $(-1, -4)$]

FINAL EXAMINATION
FORM B

1. 17
2. 3
3. 1
4. 42
5. $6/7$
6. $x = 7$
7. $x = 8$
8. $x = 13/3$
9. $x = 12$
10. $C = 5/9 (F - 32)$
11. $x > 7/3$ ———→ $7/3$
12. $x \leq 33$ ←——— 33
13. (a) $2\frac{3}{4}$ hours
    (b) 33 NICKELS
        41 DIMES
        58 QUARTERS
    (c) SHIRT = $33
        PANTS = $65
14. $-2x^3 + 8x^2 - 6x + 10$
15. $10x^2 - 3x - 18$
16. $2x^3 - 13x^2 + 30x - 27$
17. $\frac{1}{abc} - \frac{3a^2c^2}{1} + \frac{4c^2}{1}$
18. $2x^2 + 4x - 5$
19. $4xy^2(x - 2y - 4x^3y)$
20. $(x+5)(x-4)$
21. $(3x+5)(x-8)$
22. $x(2x-1)(3x+2)$
23. $16(x+2y)(x-2y)$
24. $\dfrac{2(x+2)}{5(x+3)}$
25. $-2(x-y)$
26. $\dfrac{2(x+2)}{3x}$
27. $\dfrac{-17x+29}{(x+2)(x+5)(x-5)}$
28. $\dfrac{2x+3}{5x-4}$
29. $x = 35/4$
30. $x = -8/33$
31. $x = 69/17$
32. $-6\sqrt{3}$
33. $7\sqrt{7}$
34. $3\sqrt{5}$
35. $\dfrac{3\sqrt{7}}{4}$
36. $\dfrac{2\sqrt{21}}{15}$
37. [graph: line through (0,9) and (4,1); $2x+y=9$]
38. [graph: line through (-5,0) and (2,0); $5x-2y=10$]
39. [graph: shaded region $3x+2y \geq 9$, through $(0, 9/2)$ and $(3,0)$]
40. $y = 2x - 5$
41. $y = \frac{3}{5}x - 3$
42. $(25, -11)$
43. $(-23, 16)$
44. $(0, 1)$
45. $x = 0, 4$
46. $x = 6, -6$
47. $x = 5, -3$
48. $1 \pm \sqrt{2}$
49. $x = -2/3, 1/2$
50. [graph: parabola $y = x^2 - 2x - 3$, vertex $(1, -4)$]

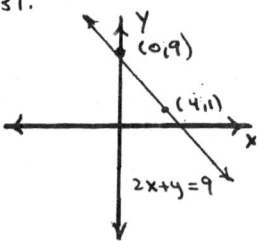

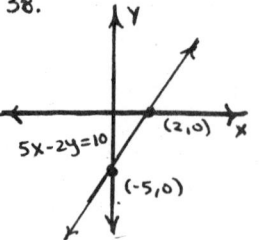

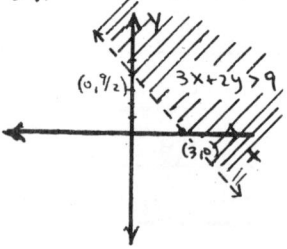

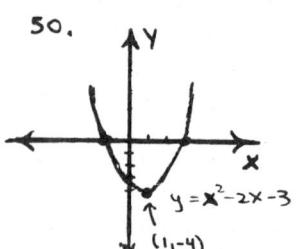

CHAPTER EXERCISES
ANSWERS

CHAPTER 1

EXERCISE 1-1 (Page 7)

1(a) $1\frac{1}{2}$ (or 1.5)   2(a) $\frac{1}{2}$ (or 0.5)   3(a) $\frac{2}{10}$ (or 0.2)
 (b) 2   (b) $1\frac{1}{4}$ (or 1.25) (b) $\frac{5}{10}$ (or 0.5)
 (c) $3\frac{1}{2}$ (or 3.5)   (c) $2\frac{3}{4}$ (or 2.75) (c) $\frac{8}{10}$ (or 0.8)
 (d) $\frac{1}{2}$ (or 0.5)   (d) $3\frac{3}{4}$ (or 3.75) (d) $1\frac{4}{10}$ (or 1.4)

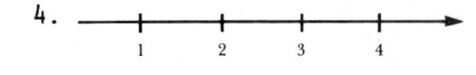

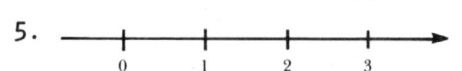

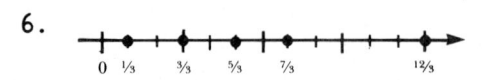

4. [number line: 1, 2, 3, 4]   5. [number line: 0, 1, 2, 3]

6. [number line: 0, ⅓, ⅗, ⁵⁄₃, ⁷⁄₃, ¹²⁄₃]   7. [number line: 0, ¼, ⅝, ⁷⁄₄, ⁸⁄₄, 2¼, 2¾]

8.  [number line: 0, ³⁄₁₀, ⁶⁄₁₀, ⁹⁄₁₀, ³⁰⁄₁₀, ³⁵⁄₁₀]

9. 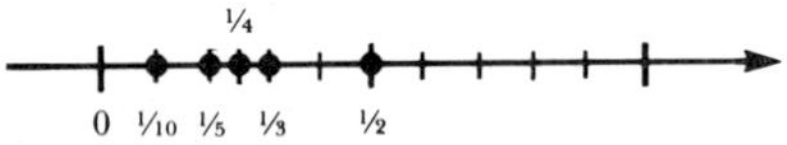 [number line: 0, ¹⁄₁₀, ⅕, ¼, ⅓, ½]

10. [number line: 0, ⅔, 0.75, 1.5, 2⅛]

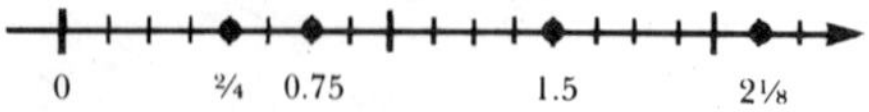

11. F, K, P, U, Z   12. A   13. All points A – Z   14. G, H, I, J, L, M, N, O, Q, R, S, T, V, W, X, Y

EXERCISE 1-2 (Page 27)

1. $\frac{1}{3}$ 2. $\frac{1}{2}$ 3. $\frac{1}{4}$ 4. $\frac{3}{5}$
5. $\frac{3}{5}$ 6. $\frac{5}{6}$ 7. $\frac{3}{2}$ 8. $\frac{3}{2}$
9. 5   10. $\frac{2}{5}$ 11. $\frac{33}{40}$ 12. $\frac{1}{125}$
13. $\frac{3}{4}$ 14. $\frac{3}{250}$ 15. $\frac{17}{20}$ 16. $1\frac{2}{5}$
17. $5\frac{3}{25}$ 18. $6\frac{12}{25}$ 19. 0.75   20. 0.2
21. 0.1   22. 0.5   23. $0.3\overline{5}$ 24. $0.\overline{8}$
25. 0.375   26. $0.8\overline{333}$ 27. $0.\overline{666}$ 28. $\frac{5}{2}$
29. $\frac{31}{6}$ 30. $\frac{23}{3}$ 31. $\frac{14}{3}$ 32. $\frac{131}{12}$
33. $\frac{83}{7}$ 34. $\frac{19}{9}$ 35. $\frac{247}{10}$ 36. $\frac{200}{3}$
37. $2\frac{1}{2}$ 38. 4   39. 1   40. 9
41. $2\frac{1}{5}$ 42. $6\frac{7}{16}$ 43. $3\frac{3}{7}$ 44. $3\frac{2}{3}$
45. $5\frac{3}{7}$ 46. 14 is less than 112

47. 3.6 is not greater than 4.2   48. $\frac{1}{2}$ is greater than $\frac{1}{8}$
49. $\frac{3}{8}$ does not equal $\frac{2}{3}$ 50. 5 is greater than 3 and less than 8
51. $\frac{3}{4}$ is greater than $\frac{1}{2}$ and less than $\frac{2}{3}$
52. 4 < 7   53. 3 > 0   54. 6 = 6
55. $\frac{5}{6}$ < $\frac{7}{6}$ 56. $\frac{8}{9}$ > $\frac{3}{6}$ 57. $\frac{3}{8}$ < $1\frac{3}{4}$
58. 0.6 < 7.2   59. 0.9 > 0.256   60. 1.36 < 1.4

EXERCISE 1-3 (Page 27)

1. $\frac{46}{35}$ 2. $\frac{10}{9}$ 3. $\frac{41}{24}$ 4. $\frac{7}{12}$
5. $\frac{5}{36}$ 6. $\frac{1}{2}$ 7. $\frac{15}{28}$ 8. $\frac{2}{1}$
9. $\frac{1}{20}$ 10. $\frac{8}{9}$ 11. $\frac{15}{4}$ 12. $\frac{3}{1}$
13. $3\frac{11}{12}$ 14. $16\frac{1}{9}$ 15. $40\frac{5}{24}$ 16. $1\frac{2}{7}$
17. $2\frac{11}{12}$ 18. $3\frac{5}{8}$ 19. $\frac{51}{8}$ 20. 6
21. 12   22. $\frac{9}{5}$ 23. $\frac{9}{10}$ 24. $\frac{6}{5}$
25. 49.9538   26. 75.481   27. 2.3   28. 593.07
29. 0.3990   30. 0.009072   31. 0.04   32. 85

EXERCISE 1-4 (Page 31)

1. 10^2 2. 4^3 3. 7^5 4. 12^4
5. 6^0 6. 2^1 7. 2^8 8. 15^3
9. 9^0 10. x^3 11. 2^3 12. 4^5
13. 10^3 14. x^5 15. y^3 16. $(3x)^3$
17. $2^3 \cdot 3^4$ 18. $4^2 \cdot 3^3 \cdot 5^5$ 19. $9^1 \cdot x^3$ 20. $10^2 \cdot x^3 \cdot y^4$
21. 16   22. 1   23. 10,000   24. 16
25. 125   26. 36   27. 48   28. 648
29. 75   30. 27   31. 40   32. 192
33. 120   34. 896   35. 168   36. 225
37. 900   38. 108   39. 96   40. 3600
41. 3240   42. 882   43. 1440   44. 95,256
45. 6   46. 810,000   47. 1000   48. $2^0 = 1$

EXERCISE 1-5 (Page 37)

1. 28   2. 38   3. 6   4. 39
5. 0   6. 5   7. 48   8. 27
9. 69   10. 16.6   11. 14   12. 9
13. 24   14. 100   15. 36   16. $\frac{1}{4}$ (or 0.25)

218

EXERCISE 1-5 (Cont'd)

17.	8	18.	336	19.	120	20.	5
21.	40	22.	492	23.	1	24.	42
25.	6	26.	1625	27.	109	28.	3
29.	42.875	30.	104,976	31.	729	32.	$\frac{1}{8}$
33.	125.81	34.	36	35.	369	36.	117
37.	$3\frac{4}{9}$	38.	$\frac{20}{49}$	39.	$\frac{1369}{3600}$	40.	$\frac{59}{4}$

REVIEW EXERCISES (Page 38)
CHAPTER 1

1.   [number line with points at 1, 2, 3, 4, 5]   2. 0

3. There are many fractions equal to the given numbers; we list three.

   (a) $2 = \frac{2}{1}$ $\frac{2 \times 2}{1 \times 2} = \boxed{\frac{4}{2}}$ $\frac{2 \times 3}{1 \times 3} = \boxed{\frac{6}{3}}$ $\frac{2 \times 4}{1 \times 4} = \boxed{\frac{8}{4}}$

   (b) $6 = \frac{6}{1}$ $\frac{6 \times 2}{1 \times 2} = \boxed{\frac{12}{2}}$ $\frac{6 \times 3}{1 \times 3} = \boxed{\frac{18}{3}}$ $\frac{6 \times 4}{1 \times 4} = \boxed{\frac{24}{4}}$

   (c) $0 = \frac{0}{1}$ $\frac{0 \times 2}{1 \times 2} = \boxed{\frac{0}{2}}$ $\frac{0 \times 3}{1 \times 3} = \boxed{\frac{0}{3}}$ $\frac{0 \times 4}{1 \times 4} = \boxed{\frac{0}{4}}$

4. There are many fractions equivalent to the given fractions; we list three.

   (a) $\frac{2}{3} = \frac{2 \times 2}{3 \times 2} = \boxed{\frac{4}{6}}$ $\frac{2 \times 3}{3 \times 3} = \boxed{\frac{6}{9}}$ $\frac{2 \times 4}{3 \times 4} = \boxed{\frac{8}{12}}$

   (b) $\frac{3}{5} = \frac{3 \times 2}{5 \times 2} = \boxed{\frac{6}{10}}$ $\frac{3 \times 3}{5 \times 3} = \boxed{\frac{9}{15}}$ $\frac{3 \times 4}{5 \times 4} = \boxed{\frac{12}{20}}$

   (c) $\frac{2}{4} = \frac{2 \div 2}{4 \div 2} = \boxed{\frac{1}{2}}$ $\frac{2 \times 2}{4 \times 2} = \boxed{\frac{4}{8}}$ $\frac{2 \times 3}{4 \times 3} = \boxed{\frac{6}{12}}$

5. (a) $\frac{3}{10}$ 6. (a) $\frac{5}{3}$ 7. (a) Point F
   (b) $\frac{7}{10}$ (b) $\frac{7}{2}$ (b) Point C
   (c) $2\frac{3}{5}$ (c) $\frac{21}{5}$ (c) Point B

8.	12.96	9.	53.21	10.	165.51	11.	15.3954
12.	259.12	13.	533.19	14.	5.714	15.	99.57
16.	28.2009	17.	$\frac{5}{6}$	18.	$\frac{5}{7}$	19.	1
20.	$\frac{5}{8}$	21.	$\frac{7}{12}$	22.	$\frac{23}{36}$	23.	$9\frac{5}{8}$
24.	$9\frac{5}{12}$	25.	$11\frac{2}{5}$	26.	1.53	27.	3.9
28.	200.064	29.	1.92	30.	5.39	31.	2.441
32.	6.13	33.	709.36	34.	4.487	35.	$\frac{1}{3}$

REVIEW EXERCISES
CHAPTER 1 (Cont'd)

36. $\frac{1}{9}$ 37. $\frac{6}{13}$ 38. $\frac{1}{2}$ 39. $\frac{5}{24}$

40. $\frac{11}{18}$ 41. $1\frac{1}{2}$ 42. $1\frac{3}{4}$ 43. $2\frac{1}{6}$

44. 3.64   45. 0.0792   46. 0.00372   47. 5.26

48. 58.7664   49. 11.2336   50. 0.0009   51. 65.12

52. 51   53. $\frac{15}{28}$ 54. $\frac{3}{10}$ 55. $\frac{3}{8}$

56. $\frac{7}{16}$ 57. $\frac{1}{4}$ 58. $\frac{2}{9}$ 59. $26\frac{2}{3}$

60. 6   61. $21\frac{1}{3}$ 62. 57.5   63. 60

64. 12.3   65. 1.4   66. 65.4   67. 3.6

68. 320   69. 10.8   70. 0.6   71. 2

72. $\frac{9}{2}$ 73. $\frac{7}{6}$ 74. $\frac{21}{4}$ 75. $\frac{9}{7}$

76. $\frac{160}{9}$ 77. $\frac{7}{9}$ 78. $\frac{5}{9}$ 79. $\frac{6}{11}$

80. (a) $13 > 5$ 81. (a) 144
    (b) $5.6 > 2.007$ (b) 1800
    (c) $\frac{3}{4} > \frac{2}{3}$ (c) 1
    (d) $0.2 < \frac{3}{5}$

82. (a) 32 (b) 45 (c) 75 (d) 54 (e) 69 (f) $\frac{1}{6}$ (g) 30
    (h) 1 (i) 73 (j) 3000 (k) 63 (l) 900

CHAPTER 2
EXERCISE 2-1 (Page 46)

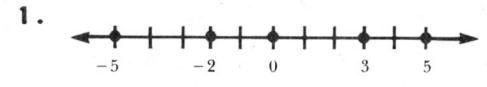

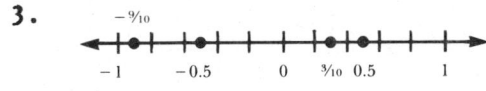

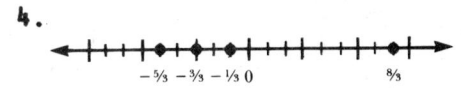

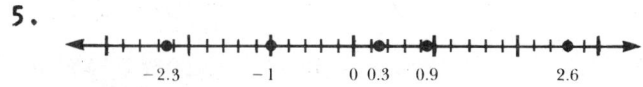

1. [number line with points at −5, −2, 0, 3, 5]
2. [number line with points at −3, −1½, −¼0, ¾, 2¼]
3. [number line with points at −⁹⁄₁₀, −1, −0.5, 0, ³⁄₁₀, 0.5, 1]
4. [number line with points at −⁵⁄₃, −³⁄₃, −¹⁄₃, 0, ⁸⁄₃]
5. [number line with points at −2.3, −1, 0, 0.3, 0.9, 2.6]

6. $-3 < +5$ 7. $0 < +6$ 8. $-4 > -5$ 9. $+1\frac{1}{2} > -\frac{3}{2}$

EXERCISE 2-1 (Cont'd)

10. $+5 > -5$ 11. $+1.2 > -\frac{8}{4}$ 12. $-3.1 > \frac{-32}{10}$ 13. $-9 < +0.9$

14. $+5 > -100$ 15. $-1\frac{3}{4} > \frac{-15}{8}$ 16. $+2\frac{3}{10} > -2.4$ 17. $-2\frac{7}{8} < -2\frac{3}{4}$

18. $-7 < -5 < -4$ 19. $-9.1 < -0.6 < 8\frac{1}{8}$

20. $-3 < -\frac{1}{2} < 2.5$

21. +$300   22. -100 feet
23. +$10   24. +2   25. -20   26. -12 pounds   27. -$100
28. -12 degrees   29. -12 yards   30. -$206   31. +6
32. +3.4   33. -0.5   34. $-\frac{9}{12}$ 35. 0   36. $+\frac{7}{3}$
37. Negative   38. Negative   39. Positive
40(a) The opposit of (+3) is represented as -(+3);
      The opposite of (+3) is -3
40(b) The opposite of (-3) would be represented as -(-3);
      The opposite of (-3) is +3

EXERCISE 2-2 (Page 50)

1. 6   2. 3   3. 8.5   4. $7\frac{1}{2}$ 5. 9.42   6. $3\frac{2}{3}$

7. $4\frac{1}{4}$ 8. $3\frac{2}{3}$ 9. 5   10. 0.9   11. $|6| > -6$

12. $|-5| = 5$ 13. $|-5| > -5$ 14. $|-3| = |-3|$ 15. $|-3| = |3|$
16. $|9| = |-9|$ 17. $|-5| > 2$ 18. $|-8| < |-12|$
19. $14 > -36$ 20. 8   21. 11   22. 27   23. 1
24. 27   25. 50   26. 2   27. 9   28. 0   29. 9

EXERCISE 2-3 (Page 60)

1. +11   2. -1   3. +4   4. 15   5. -28

6. +28   7. -4   8. +1   9. +2   10. -2

11. +2   12. -2   13. +8   14. 0   15. -9

16. -43   17. -156   18. -77   19. +2.287   20. +2.71

21. -3.92   22. -9.8   23. -1.7   24. -2.2   25. +8.85

26. -5.88   27. +1.24   28. $+\frac{1}{3}$ 29. $-\frac{8}{9}$ 30. 1

31. -1   32. $-\frac{1}{5}$ 33. $+\frac{1}{2}$ 34. +1   35. $-\frac{1}{2}$

36. -3   37. $-\frac{5}{6}$ 38. $-\frac{46}{63}$ 39. $-\frac{73}{15}$ 40. +31

41. +1   42. -16   43. -46   44. -13   45. -14

46. +14   47. 0   48. -6   49. -4   50. -45

51. +12   52. +23   53. +30   54. -8   55. -10

EXERCISE 2-3 (Cont'd)

56. +135; Ron won money. He only won $35 however, since he started with $100.

57. $-\frac{1}{2}$ 58. $-8°C$ 59. +20; 4 yards per play average

60. +15,500 feet   61. balance: $99.99; no, she is not overdrawn

62. Jim: 182; Mike: 184; Mike wins by 2 points

63. weight of freight is 54,296 pounds   64. 3,216 pounds   65. $-87°F$

EXERCISE 2-4 (Page 68)

1.	+72	2.	-28	3.	-200	4.	+48	5.	+30
6.	+100	7.	-72	8.	-16	9.	-200	10.	-51
11.	-5.4	12.	-14.4	13.	-118.3	14.	+2	15.	+461
16.	0	17.	0	18.	+1	19.	$+\frac{3}{25}$	20.	$-\frac{10}{3}$
21.	$-\frac{96}{5}$	22.	$+\frac{4}{21}$	23.	-1	24.	$-\frac{23}{24}$	25.	-240
26.	-96	27.	+100	28.	+60	29.	-60	30.	-15
31.	0	32.	+30	33.	-25	34.	-75	35.	0
36.	+9	37.	+8	38.	-8	39.	-8	40.	0
41.	-11	42.	+25	43.	+64	44.	+625	45.	+625
46.	+36	47.	+180	48.	-16	49.	-27	50.	+81
51.	-36	52.	-36	53.	-1	54.	+1	55.	-540
56.	-1800	57.	+324						

EXERCISE 2-5 (Page 73)

1.	+2	2.	-3	3.	-20	4.	+3	5.	-7
6.	+2	7.	+2	8.	+7	9.	+1	10.	-5
11.	-5	12.	0	13.	$+\frac{7}{9}$	14.	$+\frac{2}{3}$	15.	$-\frac{1}{2}$
16.	-3	17.	+2.8	18.	-4.5	19.	$-\frac{8}{9}$	20.	$-\frac{4}{3}$
21.	$+\frac{1}{18}$	22.	-2	23.	+29	24.	-43	25.	-4
26.	$-\frac{1}{3}$	27.	$+\frac{2}{3}$	28.	-30	29.	+189	30.	-27

EXERCISE 2-6 (Page 76)

1.	-4	2.	+44	3.	+6	4.	-9	5.	-36	6.	+12
7.	-178	8.	+38	9.	+2	10.	-4	11.	8	12.	-48
13.	-14	14.	-1	15.	+16	16.	+10	17.	-66	18.	-54
19.	-5	20.	6	21.	0	22.	-1	23.	+9	24.	21
25.	10	26.	-14	27.	-5	28.	-6	29.	+541	30.	$\frac{102}{25}$

EXERCISE 2-7 (Page 88)

1. True; Commutative Property Of Multiplication
2. False   3. False   4. False
5. True; Distributive Property
6. True; Identity Property Of Multiplication
7. True; Associative Property Of Multiplication
8. False   9. True; Identity Property Of Addition
10. True; Distributive Property   11. False
12. True; Commutative Property Of Addition   13. False
14. True; Distributive Property   15. False
16. False   17. True; Distributive Property
18. True; Identity Property Of Multiplication   19. False
20. True; Associative Property Of Addition
21. Commutative Property Of Addition
22. Commutative Property Of Addition   23. Additive Inverse
24. Identity Property Of Addition   25. Associative Property Of Addition
26. Commutative Property Of Multiplication
27. Zero Property Of Multiplication
28. Associative Property Of Multiplication
29. Commutative Property Of Multiplication
30. Distributibe Property

31.	(8 + 2)x	32.	6(3 - 9)	33.	(4 + 1)y
34.	z(5 - 1)	35.	5(3) + 5(4)	36.	9(12) + 9(2)
37.	12(x) - 12(y)	38.	3(4) - 5(4)	39.	8(6) + 1(6)
40.	z(x) - z(y)	41.	30	42.	-30
43.	2	44.	45	45.	-6
46.	6	47.	8	48.	-8

REVIEW EXERCISES (Page 89)
(Chapter 2)

1.	g	2.	i	3.	f	4.	h	5.	b	6.	e	7.	a	8.	c	9.	d
10.	+2	11.	-14	12.	+10	13.	-16	14.	-2								
15.	+17	16.	+12	17.	+33	18.	22	19.	-112								
20.	24	21.	-35	22.	15	23.	19	24.	-4								
25.	-18	26.	+54	27.	+10	28.	-216	29.	-1728								
30.	+144	31.	+9	32.	-1	33.	-5	34.	-2								
35.	-16	36.	-64	37.	62	38.	-5	39.	-54								
40.	56	41.	60	42.	192	43.	-150	44.	-36								
45.	0																

CHAPTER 3
EXERCISE 3-1 (Page 97)

1. (a) variable is x
   (b) constant is +6

2. (a) variable is h
   (b) constant is -8

3. (a) variable is p
   (b) constant is +15

4. (a) variables: a and b
   (b) constant is +8

5. (a) variables: x and y
   (b) constant is -16

6. (a) variables: x
   (b) constant is +5

7. (a) variables: x and y
   (b) constant is $-\frac{3}{4}$

8. (a) variables: x and y
   (b) constant is $+\frac{6}{17}$

9. (a) variables: x and y
   (b) constant is -8

10. (a) two terms
    (b) 1st term: 3x
        2nd term: +9y
    (c) coefficient of 1st term: 3
        coefficient of 2nd term: +9

11. (a) two terms
    (b) 1st term: 4y
        2nd term: -6z
    (c) coefficient of 1st term: 4
        coefficient of 2nd term: -6

12. (a) one term
    (b) term: 12xy
    (c) coefficient is 12

13. (a) three terms
    (b) 1st term: x
        2nd term: -4z
        3rd term: +16
    (c) coefficient of 1st term: 1
        coefficient of 2nd term: -4
        constant term: +16

14. (a) three terms
    (b) 1st term: 5a
        2nd term: -b
        3rd term: -8
    (c) coefficient of 1st term: 5
        coefficient of 2nd term: -1
        constant term: -8

15. (a) three terms
    (b) 1st term: x
        2nd term: -y
        3rd term: -z
    (c) coefficient of 1st term: 1
        coefficient of 2nd term: -1
        coefficient of 3rd term: -1

16. (a) three terms
    (b) 1st term: $3x^2$
        2nd term: +2x
        3rd term: -4
    (c) coefficient of 1st term: 3
        coefficient of 2nd term: +2
        constant term: -4

17. (a) two terms
    (b) 1st term: 4x
        2nd term: (3y + 6z)
    (c) coefficient of 1st term: 4
        coefficient of 2nd term: 1

18. (a) one term
    (b) the term is: (x + y + z)
    (c) coefficient of this term is 1

19. (a) one term
    (b) term is: -4x(3y + 6z)
    (c) coefficient is -4

20. (a) three terms
    (b) 1st term: $6x^3$
        2nd term: $+3x^2$
        3rd term: -4(x + y)
    (c) coefficient of 1st term: 6
        coefficient of 2nd term: +3
        coefficient of 3rd term: -4

21. (a) three terms
    (b) 1st term: $3ab^2$
        2nd term: $(+\frac{2x - y}{3xy})$
        3rd term: $+2(x^2 - 6y)$
    (c) coefficient of 1st term: 3
        coefficient of 2nd term: +1
        coefficient of 3rd term: +2

EXERCISE 3-1 (Cont'd)

22. x^2
23. $x^3 y^1$
24. $x^2 y^4$
25. $5x^2 y$
26. $-3 \cdot xy^2 z$
27. $-16x^2 y^2$
28. $3x^3 y^3 z$
29. $3(x^3 + y^2)$
30. $3(2x^3 - 4y^2)$
31. $x \cdot x$
32. $2 \cdot x \cdot x \cdot x \cdot x$
33. $-5 \cdot y \cdot y \cdot y$
34. $3xxyyyy$
35. $(-5x)(-5x)$
336. $15(-x)(-x)(-x)yyyy$

EXERCISE 3-2 (Page 102)

1. 15
2. 18
3. 1
4. 142
5. -6
6. 21
7. 9
8. -8
9. $\frac{5}{2}$
10. $\frac{19}{6}$
11. 8
12. $\frac{9}{4}$
13. 0
14. 0
15. -18
16. -18
17. -14
18. +48
19. -14
20. -20
21. -420
22. +1
23. +21
24. $+\frac{2}{5}$
25. 100
26. 64
27. 164
28. -64
29. 36
30. 16
31. -52
32. +52
33. -16
34. -12
35. 29
36. -11
37. C = 20
38. A = 160
39. A = 24
40. F = 68
41. C = 27.2
42. c = 0.43
43. S = 104
44. P.D.= 148.4406
45. S = 185.9
46. A = 155
47. A = 6655
48. S = +8
49. ℓ = 0.3125
50. $\ell = -\frac{39}{2}$

51. (a) 0.4 amps
    (b) 14 amps
    (c) 240 amps
52. (a) 1001 watts
    (b) 11,000 watts
53. (a) 2300° F
    (b) 1940° F
54. (a) 81.6°C
    (b) -20°C
55. 1800 horsepower
56. (a) 13.924
    (b) 34.848
57. (a) 90.7 cubic inches
    (b) 1397 cc
58. (a) 51 feet
    (b) 35 feet
59. (a) 400 candlepower
    (b) Yes, the intensity level for Setek's office is 70.9 candlepower.

60. 144 feet
61. Approximately 62.5%
62. 2.13 inches
63. $27,000
64. 90
65. (a) 158 beats per second
    (b) 86.6 beats per second
    (c) between 6 and 22 minutes

EXERCISE 3-3 (Page 109)

1. $11x$
2. y
3. $-94t$
4. $-11a$
5. $14m$
6. $60x$
7. $17x$
8. 0
9. $4x + 12y$
10. $20y - 3ab$
11. $9c + 8d$
12. $-5m$
13. $-2w - 3x$
14. $-15r - 18s$
15. $5(x + y)$
16. $-5(2a)$
17. $6xy - 3x$
18. $-4mn + 6n$
19. $3x^2 + 12x$
20. $14x^2 + 11x - 3xy + 2y^2$
21. $-3x^2 y$
22. $-5x^2 y^2$
23. $-3x^2 y + 8xy^2$

EXERCISE 3-3 (Cont'd)

24. $-4x^2 - x^3$
25. $-a^2b + 3ab^2$
26. $8x^3 + x^2 - 3x$
27. $-5 + 5ab + 5a - b$
28. $6xy - 13x + 3y$
29. $6xy^2 - 3x^2y^2 + 2x^2y - 6x$
30. $3x - 5y^2 - 6x^2$
31. $3x - y$
32. $6x + 5y$
33. $-x + 24$
34. $15 + x$
35. $2x - 7$
36. $-x + 9y$
37. $33x + 48$
38. $39 - 37x$
39. $6a - 11b$
40. $2x - 71$
41. $5x^2 - 13x + 84$
42. $6x^2 + 38x - 58$
43. $7x^2 - 24x + 8$
44. $-2x^2 - 7x + 9$
45. $19x^2 - 6x - 35$
46. $44x + 54$
47. $45x - 100$
48. $-a + 18$
49. $39x - 130$
50. $88a - 333$
51. $13x + 87y$
52. $-24a + 48b$
53. $-46x + 59y$
54. $6x + 72y + 24$
55. $-6x + 9a$
56. $-43x + 132y - 40$
57. $-20x - 20y$
58. $8y$
59. $230x - 285$
60. $-198x + 480$

REVIEW EXERCISES
CHAPTER 3 (Page 113)

1. variable
2. variable
3. term
4. term
5. constant
6. constant
7. coefficient
8. coefficient
9. factor
10. factor
11. coefficient
12. factor
13. -1

14 (a) five terms
14 (b) 1st term: $8x^4$
2nd term: $+5x^3$
3rd term: $-x^2$
4th term: $+7x$
5th term: -6

14 (c) coefficient of 1st term: 8
coefficient of 2nd term: +5
coefficient of 3rd term: -1
coefficient of 4th term: +7
constant term is -6

15. $5x^4y^3$
16. $8xxxyyyy$
17. 15
18. -15
19. -6
20. 9
21. -270
22. $I = 25{,}000$
23. $\ell = \frac{3}{16}$
24. $A = 50.24$
25. $F = 59$
26. $5x - y$
27. $6a - 12b$
28. $21x^2 - 30x - 4$
29. $-10x + 23y$

CHAPTER 4
EXERCISE 4-2 (Page 120)

1 (a) addition
  (b) subtract 6 from both sides of equation

2 (a) multiplication
  (b) divide both sides of equation by 2

3 (a) subtraction
  (b) add 5 to both sides of equation

4 (a) division
  (b) multiply both sides of equation by 10

5 (a) subtraction
  (b) add 5 to both sides of equation

6 (a) multiplication
  (b) divide both sides of equation by 5

EXERCISE 4-2 (Cont'd)

7(a) addition
 (b) subtract 23 from both sides of equation

8(a) division
 (b) multiply both sides of equation by 4

9(a) addition and multiplication
 (b) first subtract 5 from both sides of equation, then divide both sides of equation by 2

10.	$x = 5$	11.	$x = 3$	12.	$x = 4$	13.	$x = 17$
14.	$x = 4$	15.	$x = 25$	16.	$x = -12$	17.	$x = -25$
18.	$x = -10$	19.	$x = -7$	20.	$x = -9$	21.	$x = 19$
22.	$x = 19$	23.	$x = 15$	24.	$x = -1$	25.	$x = 22$
26.	$x = 29$	27.	$x = 2$	28.	$x = 4$	29.	$x = -28$
30.	$x = 19$	31.	$x = 8$	32.	$x = 3$	33.	$x = 5$
34.	$x = 8$	35.	$x = 27$	36.	$x = 10$	37.	$x = -9$
38.	$x = -7$	39.	$x = -3$	40.	$x = -24$	41.	$x = -24$
42.	$x = -54$	43.	$x = -8$	44.	$x = 6$	45.	$x = -2$
46.	$x = 54$	47.	$x = 60$	48.	$x = -24$	49.	$x = 1$
50.	$x = +\frac{3}{2}$	51.	$x = -9$				

EXERCISE 4-3 (Page 125)

1.	$x = 3$	2.	$x = 2$	3.	$x = 7$	4.	$x = 8$
5.	$x = 1$	6.	$x = 12\frac{1}{2}$	7.	$x = -5$	8.	$x = -8$
9.	$x = -11\frac{1}{3}$	10.	$x = -2$	11.	$x = -3$	12.	$x = -5$
13.	$x = -2$	14.	$x = -1$	15.	$x = -25$	16.	$x = -3$
17.	$x = -22$	18.	$x = 0$	19.	$x = 10$	20.	$x = 15$
21.	$x = 15$	22.	$x = 48$	23.	$x = -30$	24.	$x = 42$
25.	$x = 15$	26.	$x = 20$	27.	$x = -28$	28.	$x = 25$
29.	$x = -54$	30.	$x = 15$	31.	$x = 56$	32.	$x = -7$
33.	$x = 80$	34.	$x = -21$	35.	$x = 21$	36.	$x = -18$
37.	$x = -9$	38.	$x = 27$	39.	$x = 5$	40.	$x = -15$
41.	$x = 12$	42.	$x = -14$	43.	$x = -5$	44.	$x = -31$
45.	$x = -13$	46.	$x = -12$	47.	$x = -30$	48.	$x = 14$
49.	$x = -12$	50.	$x = -12$	51.	$x = -20$		

EXERCISE 4-4 (Page 128)

1.	$x = 3$	2.	$x = 2$	3.	$x = 3$	4.	$x = 2$
5.	$x = -72$	6.	$y = -4$	7.	$x = 31$	8.	$x = 11$
9.	$x = -60$	10.	$x = 12$	11.	$x = 2$	12.	$x = -6.8$
13.	$x = -8$	14.	$x = -21$	15.	$x = 2$	16.	$x = -5$
17.	$x = 100$	18.	$x = 5$	19.	$x = 275/2$	20.	$x = -385$

EXERCISE 4-5 (Page 132)

1.	$x = 5$	2.	$x = 2$	3.	$x = 7$	4.	$x = 16$
5.	$x = -5$	6.	$x = -3$	7.	$x = 7$	8.	$x = 4$
9.	$x = -2$	10.	$x = 5$	11.	$x = -18$	12.	$x = 1$
13.	$x = 1$	14.	$x = -16$	15.	$x = \frac{6}{11}$	16.	$x = 1$
17.	$x = 3$	18.	$x = 10$	19.	$x = 3$	20.	$x = 8$

EXERCISE 4-6 (Page 136)

1. $x = 3$	2. $x = 8$	3. $x = 8$	4. $x = 1$
5. $x = 1$	6. $x = -4$	7. $x = 2$	8. $x = 8$
9. $x = 9$	10. $x = -5$	11. $x = -7$	12. $x = -1$
13. $x = -10$	14. $x = 6$	15. $x = 23$	16. $x = -8$
17. $x = -11$	18. $x = -\frac{26}{3}$	19. $x = 2$	20. $x = 6$
21. $x = -5$	22. $x = -5$	23. $x = 2$	24. $x = 2$
25. $x = 4$	26. $x = 10$	27. $x = -12$	28. $x = -35/19$
29. $x = 3$	30. $x = -\frac{49}{8}$	31. $x = -\frac{40}{19}$	32. $x = 9$
33. $x = -7$	34. $x = \frac{28}{13}$	35. $x = \frac{28}{5}$	36. $x = \frac{26}{7}$
37. $x = \frac{11}{9}$	38. $x = 0$	39. $x = 2$	40. $x = \frac{18}{11}$
41. $x = \frac{24}{5}$	42. $x = -2$	43. $x = \frac{30}{7}$	44. $x = 1$
45. $x = 3$	46. $x = 45$	47. $x = -5$	48. $x = 0$
49. $x = 12$	50. $x = 5$	51. $x = -\frac{19}{6}$	52. $x = 2$
53. $x = 80$	54. $x = 4$		

EXERCISE 4-7 (Page 139)

1. $H = 19 - \ell$
2. $B = A - 3$
3. $\frac{F}{A} = M$
4. $\frac{D}{60} = t$
5. $\frac{C}{D} = \pi$
6. $H = \frac{A}{B}$
7. $W = \frac{V}{LH}$
8. $r = \frac{C}{2\pi}$
9. $b = y - mx$
10. $y = \frac{-C - Ax}{B}$
11. $b = \frac{2A}{H}$
12. $g = \frac{3p}{MH}$
13. $C = \frac{5}{9}(F - 32)$
14. $N = 4(F - 40)$
15. $a = \frac{2s}{N} - \ell$
16. $C = \frac{2a}{h} - b$
17. $F = \frac{9}{5}C + 32$
18. $W = \frac{P}{2} - \ell$
19. $t = \frac{A}{pr} - \frac{1}{r}$
20. $d = \frac{2\ell}{N(N - 1)} - \frac{2a}{(N - 1)}$

EXERCISE 4-8 (Page 149)

1. [number line showing $x < 7$, open circle at 7, shaded to the left, from 0 to 7 marked]

2. [number line showing $x \geq -1$, closed circle at -1, shaded to the right, from -6 to -1 marked]

3. $x \leq 2$

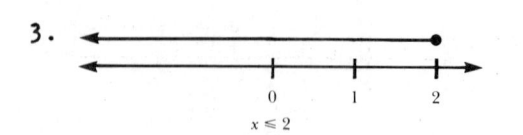

4. $x < -4$

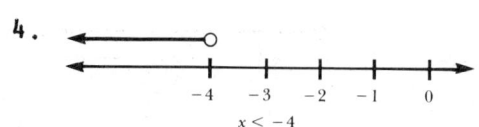

5. $x < 3$

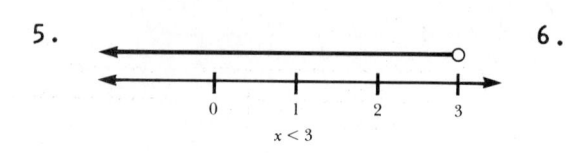

6. $x \geq 2$

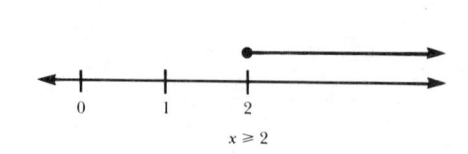

7. $x > -2$

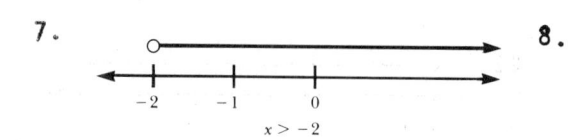

8. $x > 10$

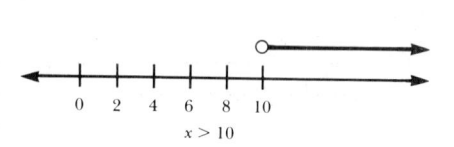

9. $x > 6$

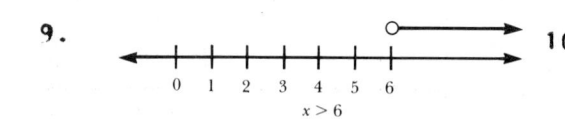

10. $x \leq -25$

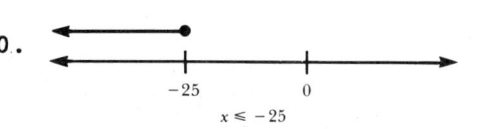

11. $x < 5$

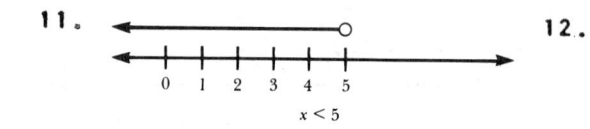

12. $x > 16$

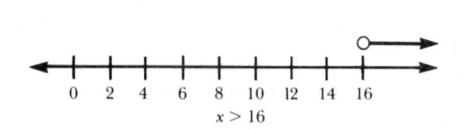

13. $x \leq 7$

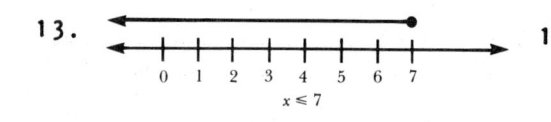

14. $x < 4$

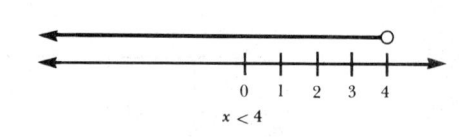

15. $x < 1$

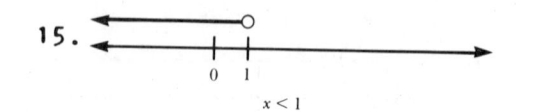

16. $x \leq 1$

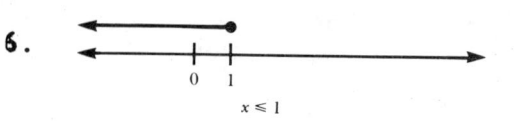

17. $x > 6/11$

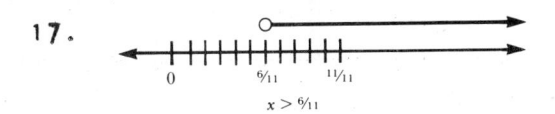

18. $x \geq 10$

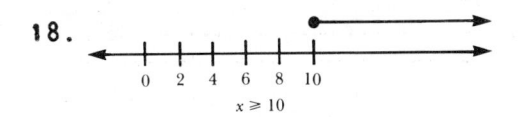

19. $x \geq 7\frac{1}{2}$

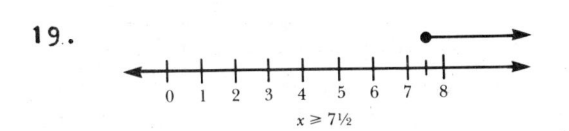

20. $x < 3$

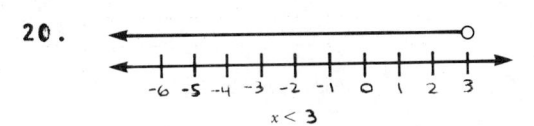

21. $x \geq -4$

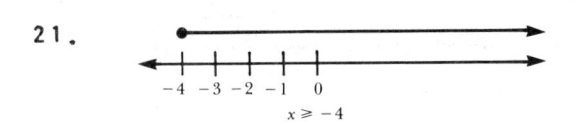

22. $x \leq 4$

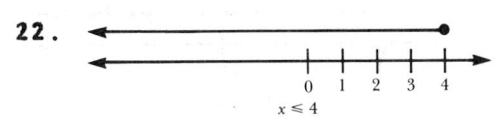

23. $x < -7$

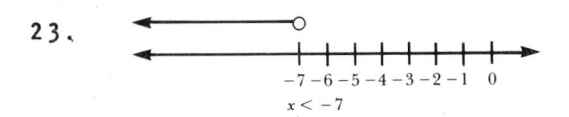

24. $x > 7/6$

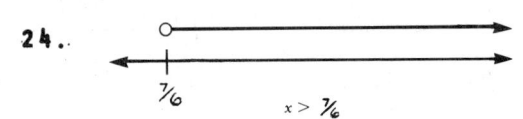

25. $x > -18/19$

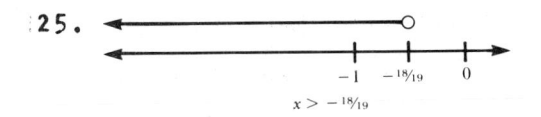

26. $x \leq 2/3$

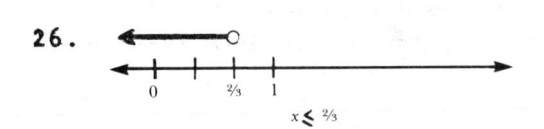

27. $x \geq 32\frac{1}{2}$

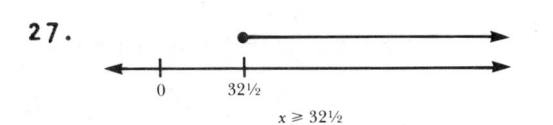

28. $x > -12$

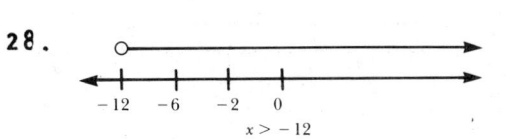

29. $x < 1$

30. $x \geq -10$

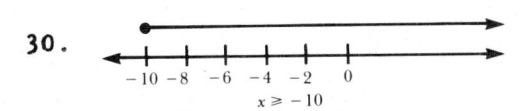

230

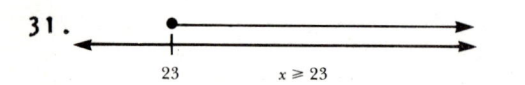

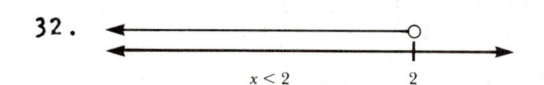

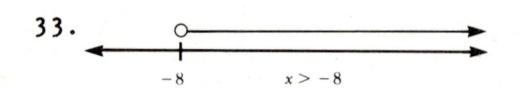

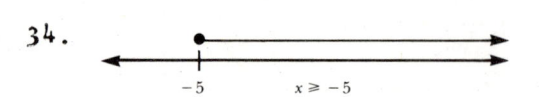

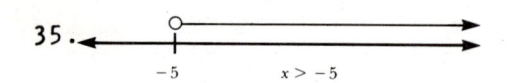

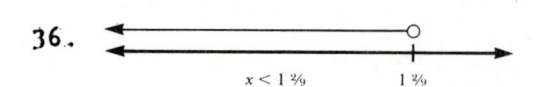

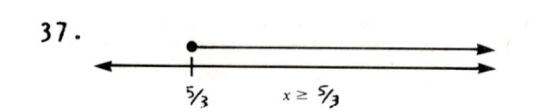

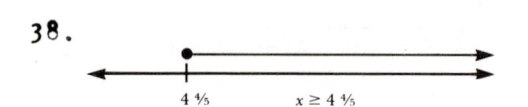

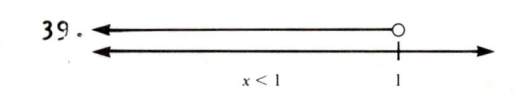

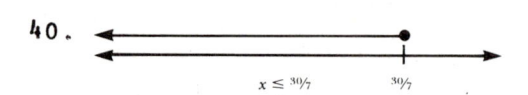

31. [number line: closed dot at 23, arrow right] $x \geq 23$

32. [number line: open dot at 2, arrow left] $x < 2$

33. [number line: open dot at -8, arrow right] $x > -8$

34. [number line: closed dot at -5, arrow right] $x \geq -5$

35. [number line: open dot at -5, arrow right] $x > -5$

36. [number line: open dot at 1⅔, arrow left] $x < 1\tfrac{2}{9}$

37. [number line: closed dot at 5/3, arrow right] $x \geq \tfrac{5}{3}$

38. [number line: closed dot at 4⅕, arrow right] $x \geq 4\tfrac{1}{5}$

39. [number line: open dot at 1, arrow left] $x < 1$

40. [number line: closed dot at 30/7, arrow left] $x \leq \tfrac{30}{7}$

REVIEW EXERCISES (Page 150)
CHAPTER 4

1. $x=3$ 2. $x=9$ 3. $x=1$ 4. $x=5$ 5. $x=-14$
6. $x=-18$ 7. $x=5$ 8. $x=16/3$ 9. $x=6$ 10. $x=19$
11. $x=-14/5$ 12. $x=6$ 13. $x=80$ 14. $x=36$ 15. $x=32$
16. $x=14$ 17. $x=-9/2$ 18. $x=-66/5$ 19. $x=-1$ 20. $x=-5/2$
21. $x=-1$ 22. $x=-14/3$ 23. $x=3$ 24. $x=3$ 25. $x=-5/3$
26. $x=1$ 27. $x=6$ 28. $x=14/3$ 29. $x=17/2$ 30. $x=-7$
31. $x=5$ 32. $x=3/2$ 33. $x=13/5$ 34. $x=45/2$ 35. $x=10$
36. $x=36$ 37. $x=-5$ 38. $x=-80/3$ 39. $x=-7$ 40. $x=28$
41. $x=5$ 42. $x=-5/2$ 43. $x=24/5$ 44. $x=-11/15$ 45. $x=-1/5$
46. $x=2$ 47. $x=7/8$ 48. $x=1$ 49. $x=-8$ 50. $x=1/11$
51. $x=19/7$ 52. $x=4$ 53. $x=-3/2$ 54. $x=-3$ 55. $x=-2$
56. $x=-2$ 57. $x=5/3$ 58. $x=1$ 59. $x=6$ 60. $x=10/3$
61. $x=4$ 62. $x=-19/4$ 63. $x=6$ 64. $x=-1$ 65. $x=4$
66. $x=15/3$ 67. $x=-1/3$ 68. $x=-5/7$ 69. $x=9/5$ 70. $x=58/5$
71. $x=-5$

REVIEW EXERCISES
CHAPTER 4 (Cont'd)

72. 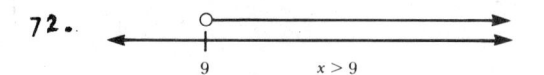 ○ at 9, $x > 9$

73. 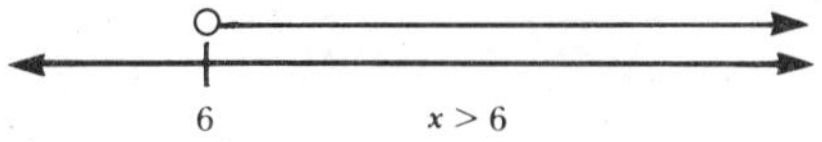 ○ at 6, $x > 6$

74. 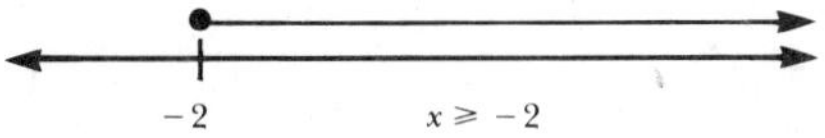 ● at −2, $x \geq -2$

75. 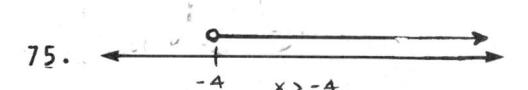 ○ at −4, $x > -4$

76. 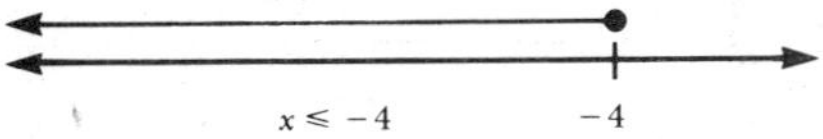 ● at −4, $x \leq -4$

77. 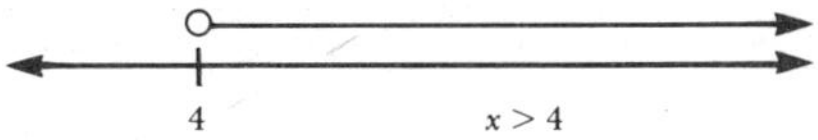 ○ at 4, $x > 4$

78. 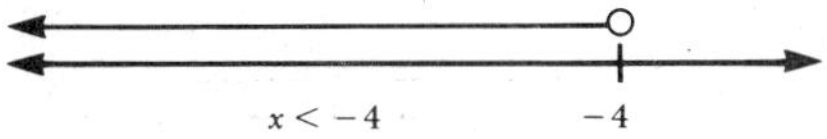 ○ at −4, $x < -4$

79. 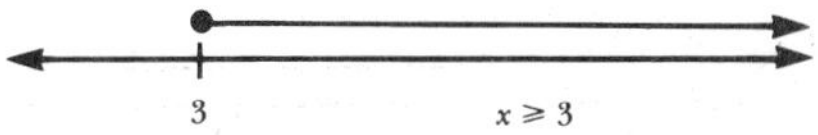 ● at 3, $x \geq 3$

80. 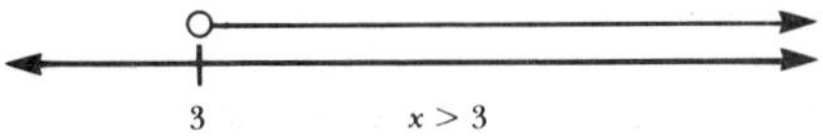 ○ at 3, $x > 3$

81. 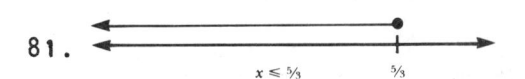 ● at 5/3, $x \leq 5/3$

82. 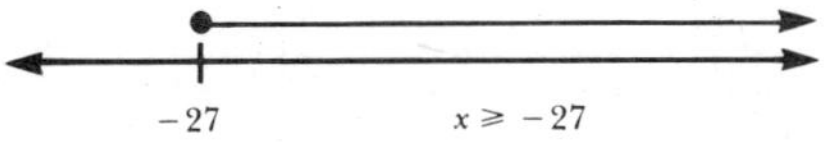 ● at −27, $x \geq -27$

83. 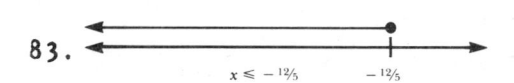 ● at −12/5, $x \leq -12/5$

84. 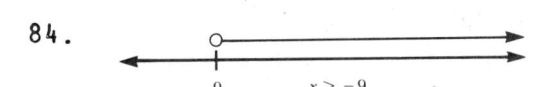 ○ at −9, $x > -9$

85. 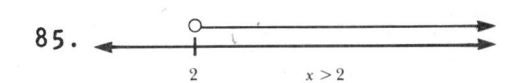 ○ at 2, $x > 2$

86. $V = \dfrac{375H}{D}$

87. $g = \dfrac{2s}{t^2}$

88. $\ell = \dfrac{2s}{N} - a$

89. $N = \dfrac{2.5\ HP}{D^2}$

CHAPTER 5
EXERCISE 5-1 (Page 157)

1. $3 + x$
2. $x - 6$
3. $x + y$
4. $x + x$ or $2x$
5. $3x - 2$
6. $(12 \div 2)x$ or $(\frac{12}{2})x$
7. $12 \div (2x)$ or $\frac{12}{2x}$
8. $x - 4y$
9. $9x$
10. $12 - 9x$
11. $2(3 + x)$
12. $a - b$
13. $x + y - (2x + 4)$
14. $x + y - 2x + 4$
15. $x \div y$ or $\frac{x}{y}$
16. $5x^2$
17. $(x - y) \div 2$ or $\frac{(x - y)}{2}$
18. $x(5 - y)$
19. $5\left[(k - 3) \div (x - y)\right]$ or $5\left(\frac{k - 3}{x - y}\right)$
20. $(x \div y)^2$ or $(\frac{x}{y})^2$
21. $x^2 \div y^2$ or $\frac{x^2}{y^2}$
22. $x \div y^2$ or $\frac{x}{y^2}$

23. $3 + x = 23$
24. $27 + 2x = 48$
25. $6x - 5 = 41$
26. $\frac{x}{6} = 3$
27. $3x - 19 = 8$
28. $3x - 1 = 17$
29. $9 + 6x = 63 + 2x$
30. $5 - 2x = 15 - x$
31. $x - y = 5x$
32. $\frac{1}{2}x = 9$
33. $18 + 2x = \frac{1}{3}x$
34. $3x - 3 = 5x + 8$
35. $\frac{x}{9} = 3x - 5$
36. $x + 6 = 38$
37. $2x - 13 = 29$
38. $x + 10 = 12$
39. $x - 5 = 23$
40. $27 = 9x$
41. $2x - 9 = 31$
42. $\frac{x}{3} = 20$
43. $2(x + 5) = 80$
44. $\frac{16 - x}{4} = 5$
45. $\frac{x}{6} + 23 = 7$
46. $\frac{6}{x + 23} = 5$
47. $\frac{4 + 3x}{8} = 15$
48. $4(8 + 2x) = 20$
49. $\frac{8 + 3x}{4} = 20$
50. $2(8 - 2x) = 14$
51. $4(12 + 3x) = 96$
52. $\frac{3x - 12}{3} = 10$

EXERCISE 5-2 (Page 163)

1. $x = 2$
2. $x = -2$
3. $x = 25$
4. $x = -6$
5. $x = 30$
6. $x = -22$
7. $x = 13$
8. $x = -17$
9. $x = \frac{44}{3}$
10. $x = 8$
11. $x = -6$
12. $x = 32$
13. First number is 12; second number is 24
14. $x = -3$
15. First number is 6; second number is 24
16. $x = -9$
17. $x = 3$
18. $x = 3$
19. $x = 15$
20. $x = -3$
21. $x = 13\frac{1}{3}$
22. $x = 9$
23. $x = 5$

EXERCISE 5-3 (Cont'd)

24. First number is 5; second number is 13
25. First number is 4; second number is 16
26. First number is 5; second number is 0; third number is 3
27. Mike's age is 28; Milt's age is 40
28. One shirt costs $18; one pair of pants costs $40

EXERCISE 5-4 (Page 170)

1. 6 nickels
   6 dimes
2. 11 nickels
   6 dimes
3. 8 dimes
   24 quarters
4. 8 quarters
   40 nickels
5. 32 nickels
   48 dimes
   16 quarters
6. 135 nickels
   27 dimes
   42 quarters
7. 37 nickels
   49 dimes
   72 quarters
8. 50 nickels
   100 dimes
   150 quarters
9. 2500 tickets at $8.50 each
   2500 tickets at $6.50 each
   5000 tickets at $4.50 each
10. 312 matinee
    512 children
    1560 adult
11. 23 fifteen cent stamps
    37 twenty-five cent stamps
12. 49 twenty cent stamps
    31 thirty cent stamps
13. 28 twenty-eight cent stamps
    58 thirty-five cent stamps
    14 forty cent stamps
14. 24 ten cent stamps
    18 twenty cent stamps
    3 fifty cent stamps

EXERCISE 5-5 (Page 175)

1. 25 lb. of sunflower seeds
   25 lb. of wheat germ nuggets
2. 20 lb. of macaroni
   10 lb. of beans
3. 260 gallons of regular unleaded
   140 gallons of super unleaded
4. 200 gallons of 10w30 oil
   100 gallons of 10w40 oil
5. (a) 40 lb. of cashews; (b) yes
6. 3 lb. of veal
7. 16 oz. of Hazel's special grass seed mix
   32 oz. of Crossman's grass seed mix
8. 80 oz. of Apple Drink (19%)
   48 oz. of Apple Tang
9. 3.2 oz. of Wheatmate
   8 oz. of Cornhusk
10. 5.5 ounces of the 80 proof Turkey
11. 8 oz. of water
12. 2 quarts
13. 4 quarts
14. 30 dozen roses
    70 dozen carnations

EXERCISE 5-6 (Page 181)

1. $350 at 6%
   $650 at 10%
2. $7000 at 5%
   $18000 at 12%
3. $320 at 5%
   $820 at 7%
   A Total of $1,140 was invested
4. $900 at $5\frac{3}{4}$%
   $2700 at 14%
   A total of $3600 was invested
5. $3000 at 6% (Savings Account)
   $500 at 18% (Stock)
6. $3200 at 12% (Treasury Certificates)
   $2700 at $8\frac{1}{2}$% (Growth Stock)
7. (a) $1300 at 8%
       $2300 at 15%
       $1000 at 12%
   (b) $4600 Total
8. $800 at 14%
   $2100 at 10%
9. $2400 at 6%
   $3600 at 11%
10. $500 at $6\frac{1}{2}$%
    $2000 at 9%
11. $2250 at 14%
    $750 at 6%
12. $900 at 9%
    $600 at $6\frac{1}{2}$%
13. $350 at 8%
    $450 at 6%
14. $1500 at 7%
    $3500 at 2%

EXERCISE 5-7 (Page 188)

1. 4 hours
2. 0.49 hr. or approximately 30 min.
3. First truck: 60 mph
   Second truck: 85 mph
4. Airplane: 125 mph
   Helicopter: 87 mph
5. (a) 6 hours
   (b) 55 miles from Howard's house
6. (a) 3 hours
   (b) yes, by 10 miles
7. (a) 3 minutes
   (b) 2.2 miles
8. (a) 3 minutes
   (b) $\frac{1}{2}$ mile from store
9. First Part: 65 mph
   Second Part: 20 mph
10. Taxi: 2.4 mph
    Train: 28 mph
11. Start: 4 A.M.
    End: 2 P.M.
12. Start: 6 A.M.
    End: 11 P.M.
13. 66 miles
14. 4.2 miles from garage
15. 4994 feet

REVIEW EXERCISES (Page 189)
CHAPTER 5

1. $x = 13$
2. $x = -3$
3. $x = -256$
4. $x = 3$
5. $x = 29$
6. $x = -6$
7. $x = -8$
8. $x = 5$
9. First number is 100; second number is 25
10. $x = 50$
11. $x = -1$
12. $x = -7$
13. $x = 2$

REVIEW EXERCISES
CHAPTER 5 (Cont'd)

14. Land: $34,500
    House: $103,500

15. 12 five dollar bills
    36 ten dollar bills
    60 twenty dollar bills

16. 28 pounds of apricots
    22 pounds of blanched nuts

17. $61,250 at 16%
    $38,750 at 12%

18. (a) 10 hours
    (b) 3 A.M.

19. 3.5 gallons

20. 34 fifteen cent stamps
    1360 eighteen cent stamps
    3400 three cent stamps

21. Insulation: $2300
    Siding: $5750

22. $450 at 3%
    $1650 at 8%
    $2100 total

23. Ken: 45 MPH
    Jacki: 60 MPH

CHAPTER 6
EXERCISE 6-1 (Page 193)

1. 2 terms; 1st degree polynomial

2. 1 term; 2nd degree polynomial

3. 3 terms; 2nd degree polynomial

4. 1 term; 1st degree polynomial

5. 2 terms; 2nd degree polynomial

6. 2 terms; 2nd degree polynomial

7. 3 terms; 1st degree polynomial

8. 2 terms; 1st degree polynomial

9. 2 terms; 1st degree polynomial

EXERCISE 6-2 (Page 197)

1. $17x$
2. $+4y^2$
3. $-5ab$
4. $-4d$
5. $-7xy$
6. $-28z^3$
7. $-32x^2y$
8. $+11abc$
9. $-19y$
10. $9(a + b)$
11. $14c$
12. $+9d$
13. $-3x^2$
14. $0c$ or 0
15. $-3(x + y)$
16. $15x$
17. $-12y$
18. $13c$
19. $+6x^2$
20. $-7xy$
21. 0
22. $3c$
23. $-5xy$
24. 0
25. $-4c^2$
26. $7a + 11$
27. $7x + 14y$
28. $5x + 2y$
29. $5a$
30. $5a - 3b - c$

EXERCISE 6-2 (Cont'd)

31. $2x^3 - 1$
32. 0
33. $2x^2 + 2xy + y^2$
34. $6a - 5b + c$
35. $2a + 17b + 4c$
36. $7y^2 - 10$
37. $8x^2 - 3x + 1$
38. $2y^2 - y + 1$
39. $3z^2 + 2z$
40. $3a^2 + 4a$
41. $3x - 6y$
42. $3x + 2y$
43. $7c - 3d$
44. -6
45. $3x^2 + x$
46. $7x^2 - 8x$
47. $10x^2 - 3x + 1$
48. $6x^2 - x - 10$
49. $-y^2 + 10y + 13$
50. $5a^2 + ab + 3b^2$
51. $-6x^2 - 4x + 8$
52. $4z^2 + 2z$
53. $+3y^2 - 3y + 2$
54. $2x^2 + 6x - 5$
55. 0

EXERCISE 6-3 (Page 202)

1. $7x$
2. $16x$
3. $+3x$
4. $-29y$
5. $0z = 0$
6. $+4x$
7. $3y$
8. $8m + 6n$
9. $1.0x^2$
10. $\frac{4}{2}A = 2A$
11. $-7a$
12. $-3x^2$
13. $3y^2$
14. $4ab$
15. $-6ab$
16. $2ab$
17. $19y$
18. $9y$
19. $0ab$ or 0
20. $8y^2$
21. $-4z$
22. $-13x^2$
23. $-11y^2$
24. $0a = 0$
25. $-4a$
26. $+5a$
27. $10(a + b)$
28. $-2(x + y)$
29. $-2.2x$
30. $+1.4cd$
31. $12.0d$
32. $5x^2$
33. $-3ab$
34. 0
35. $-25x^2$
36. $11abc$
37. 0
38. $7a$
39. $8x$
40. $18y$
41. $-8x^2$
42. $-11xy$
43. $-4(a + b)$
44. $2x^2$
45. $-7ab$
46. $3x$
47. $13xy$
48. $8a$
49. $-6x^2$
50. $8mn$
51. $-12cd$
52. $-7x^2$
53. $-4\frac{1}{2}x$
54. $7.1y^2$
55. $13x^2 - 3x$
56. $-2xy + x$
57. $-3a + 6b + 7$
58. $4x + 1$
59. $a - 5b$
60. $10x^2 - 3$
61. $-16x + 23y$
62. $16x^2 - 18x$
63. $4y - 5x - 3$
64. $8a + b + 2$
65. $4x^3 + 10x^2 + 12$
66. $2y^2 + 12y + 15$
67. $7a + 2b$
68. $-3x + 12y$
69. $-3x$
70. $-x^2 + 4$
71. $-3x^2$
72. $-2x^2 - 7x + 8$
73. $4x^2 + 2x$
74. $3x^2 - 5x + 14$
75. $8y - 4$
76. $2x^2 + 2x + 19$
77. $6x^2 - 18x + 5$
78. $x^3 - 4x^2 - 4x$
79. $a + 2b$
80. $11a - 12b$
81. $-3a + 2b$
82. $8x^2 - 10x$
83. $-17a + 13b$
84. $-4x^2 + 9x$

EXERCISE 6-3 (Cont'd)

85. $2x^2 - 6x$
86. $3x + 9$
87. $3x^2 - 24$
88. $6x^2 - 7$
89. $9x^2 + 6x + y^2 + 5y - 8$
90. $-2y + 6$
91. $-3y + 2$
92. $12x - 2$
93. 0
94. $x + 17y - 11z$
95. $14r - 18s + 9t$
96. $-6a^2 - 4a + 1$
97. $2x^2 - 3x + 1$
98. $x^3 - 3x^2 - 4x + 1$
99. $-3y^2 - 8y + 11$
100. $2y^2 + 10y - 9$

EXERCISE 6-4 Page 209)

1. a^5
2. x^5
3. y^6
4. d^3
5. a^6
6. p^8
7. a^4b^5
8. x^4y
9. c^6d^4
10. x^5y^4
11. 4^5 or 1024
12. $2^3 \cdot 5$ or 40
13. 3^3x^3 or $27x^3$
14. 3^2x^3 or $9x^3$
15. x^{2+b}
16. x^6
17. y^8
18. z^8
19. y^{10}
20. 2^6 or 64
21. 2^{15} or $32,768$
22. 3^{20}
23. $3^{m \cdot n}$
24. a^6b^4
25. z^{14}
26. y^{12}
27. a^{5b}
28. a^3b^6
29. x^8y^{12}
30. $a^4b^6c^8$
31. $49a^2$
32. $9a^2$
33. $8a^6$
34. $-27c^6d^3$
35. $4x^7$
36. $a^4b^2c^6$
37. $-27y^3z^6$

EXERCISE 6-5 (Page 217)

1. $-24abc$
2. $+6x^5$
3. $24x^3y^6$
4. $21a^3b$
5. $-4x^3y^3$
6. $-12a^3d^4$
7. $4a$
8. $-12x^2$
9. $-8y^2$
10. $2z^4$
11. $+48ab$
12. $9x^2$
13. $-6xyz$
14. $+60y^{11}$
15. $+1x^6$
16. $-72z^8$
17. $-18x^7y^5$
18. $-6a^8b^9$
19. $-20x^2y^4$
20. $36a^2$
21. $4x^2$
22. $36x^6$
23. $9a^4b^6$
24. $-8a^6b^3$
25. $27y^2$
26. $8x^4$
27. $2y^{11}$
28. $0.6x^2$
29. $4a + 4b$
30. $2x^3 - 2x^2y$
31. $3x + 6y$
32. $x^2y + x^2y^2$
33. $-3a^3 + 3a^2$
34. $6x^2 + 9xy - 12xz$
35. $2x^4 - 3x^3 + 4x^2$
36. $-x^4 - 2x^3 + 5x^2$
37. $3a^3b + 4a^2b^2 + 5ab^3$
38. $+28x^2y^3 - 14x^3y^2$
39. $3y - 6y^2 + 12y^3$
40. $3x^4y^2 + 3x^3y^3 + 3x^2y^4$
41. $10x^4y + 25x^3y^2 - 20x^2y^3$

EXERCISE 6-5 (Cont'd)

42. $8y^2 - 12y + 10$
43. $2x^2 - 4x - 7$
44. $-75x^2yz^2 + 45x^2y^2z + 15xy^2z^2$
45. $6x^3 - 12x^2 + 15x$
46. $-d^5 + d^4 - 20d^2$
47. $-12a^2bc + 6ab^2c + 3abc^2 - 12abc$
48. $-8c^2x^2 + 12c^2x + 20c^2$
49. $x^2 + 7x + 12$
50. $y^2 + 7y + 6$
51. $z^2 + 4z + 4$
52. $x^2 - 7x + 6$
53. $x^2 - 9x + 14$
54. $x^2 - 4x + 4$
55. $x^2 + 4x - 21$
56. $y^2 + 6y - 16$
57. $z^2 - 7z - 18$
58. $x^2 - 5x - 14$
59. $y^2 - 4$
60. $z^2 - 16$
61. $2a^2 + 5a + 3$
62. $3a^2 - 7a + 4$
63. $3x^2 + 2a - 8$
64. $2x^2 - 7x - 49$
65. $2y^2 - y - 1$
66. $3z^2 + 2z - 1$
67. $3 + 2z - z^2$
68. $2 - 4z + 2z^2$
69. $2x^2 - 11x - 6$
70. $2z^2 + z - 6$
71. $6a^2 + 21a + 9$
72. $12x^2 - 7x - 12$
73. $10x^2 - 6x - 4$
74. $4x^2 - 9$
75. $10x^2 - 25x + 15$
76. $6z^2 + 2z - 4$
77. $-x^2 + 6x - 8$
78. $3x^2 - 12x + 9$
79. $2x^2 + 3xy + y^2$
80. $6a^2 - ab - b^2$
81. $a^4 + a^2 - 2$
82. $x^4 - 1$
83. $x^2 + 2xy + y^2$
84. $x^2 - y^2$
85. $x^2 - 2xy + y^2$
86. $2x^2 - 3xy - 2y^2$
87. $6x^2 + 7xy - 20y^2$
88. $x^4 + 2x^2y^2 + y^4$
89. $a^2 + 2a + 1$
90. $x^2 - 4x + 4$
91. $4a^2 + 4ab + b^2$
92. $4a^2 - 12a + 9$
93. $9a^2 + 12ab + 4b^2$
94. $x^3 + 6x^2 + 12x + 8$
95. $y^3 - 9y^2 + 27y - 27$
96. $a^3 - 6a^2b + 12ab^2 - 8b^3$
97. $6a^3 + 23a^2 + 25a + 6$
98. $x^3 + 2x^2 - 3x$
99. $6x^3 + 13x^2 - 19x - 12$
100. $x^3 + 3x^2 + 3x + 1$
101. $2x^3 - 10x^2 + 6x + 12$
102. $2y^3 - 15y^2 + 22y - 24$
103. $4x^3 - 13xy^2 + 6y^3$
104. $21 - 26x + 14x^2 - 4x^3$
105. $3z^3 - 13z^2 + 15z - 4$
106. $-x^3 + 2x^2 + 11x - 12$
107. $-x^3 + 3x^2 + 13x - 15$
108. $x^4 + 3x^3 + 2x^2 + x - 1$
109. $x^4 + 2x^2 + 9$
110. $z^4 - 2z^3 - 5z^2 + 6z + 9$

EXERCISE 6-6 (Page 226)

1. x^7
2. y^3
3. x^5
4. y
5. z^{2b}
6. w
7. 10^3
8. 3^3
9. y
10. x^5
11. y^3
12. z
13. 10^2
14. ab^{b-1}
15. a^2b
16. x^2y
17. x^2yz^4
18. abc^3
19. a^2b^2
20. a^3b
21. x^2yz^7
22. 1
23. 1
24. 1
25. 1
26. 1
27. 1
28. 1

EXERCISE 6-6 (Cont'd)

29. 1
30. 1
31. x
32. c^2
33. b
34. x^5
35. ac^2
36. c
37. $\dfrac{1}{y^6}$
38. $\dfrac{1}{8^3}$
39. $\dfrac{1}{z}$
40. $\dfrac{1}{b}$
41. $\dfrac{1}{b^2}$
42. $\dfrac{1}{ab}$
43. $\dfrac{1}{abc}$
44. $\dfrac{1}{bc^2}$
45. $\dfrac{a}{b^2}$
46. 1
47. $\dfrac{x^4}{z^4}$
48. $\dfrac{a^4}{b^4}$

EXERCISE 6-7 (PAGE 235)

1. $3x^3$
2. $-5x^3y$
3. $-18ab$
4. $4b^3c^2$
5. $2x$
6. $-2x$
7. $-2y$
8. $4z$
9. b
10. $2xy$
11. $-3xy^3$
12. $-10a$
13. $2b$
14. 1
15. -1
16. b
17. $4d$
18. 2
19. $-2x^4yz^3$
20. $8xy^2$
21. $-z$
22. $3s$
23. $\dfrac{2}{r^2}$
24. $\dfrac{-6}{y}$
25. $\dfrac{-6x^3}{y^3}$
26. $\dfrac{-c}{b^2}$
27. $\dfrac{4}{y}$
28. $\dfrac{y}{6}$
29. $-5(a + b)$
30. 3
31. $2x + 3y$
32. $6x^2 - 4x$
33. $3x - 1$
34. $-4a + 2$
35. $2d^2 + d$
36. $x - 1$
37. $y + z$
38. $x^2 - x$
39. $x + 6$
40. $-z + 6$
41. $4y^2 - 3y$
42. $-3y + 2$
43. $y - 2$
44. $2y^2 + 1$
45. $-3a + 1$
46. $4a - 3b$
47. $3c - d$
48. $-3a^2 - 2a$
49. $-x^3 + 3x^2 + 2$
50. $4z^2 - 2z + 1$
51. $3x - 1 + 2y$
52. $r + h$
53. $x - 2$
54. $x - 2$
55. $x + 3$
56. $2x + 1$
57. $x + 3$
58. $x + 4$
59. $x - 5$
60. $x - 5$
61. $x - 7$
62. $2x - 3$
63. $x - 5$
64. $2x + 3$
65. $5x - 7$
66. $3x + 2$
67. $x^2 + 11$ R 27
68. $2y^2 - 7y + 6$
69. $3y^2 + 2y - 7$
70. $2y^2 - 5y + 4$
71. $3y^2 + y + 3$
72. $9x^2 + 3x + 1$
73. $4x^2 - 3x + 5$ R -6
74. $10x - 7$
75. $3x - 2$ R -9
76. $x^4 + 2x^2 + 4$
77. $2y^2 - 2y + 1$
78. $2y + 3$
79. $x^2 - 2$
80. $x - 5$ R 50

REVIEW EXERCISES (Page 237)
CHAPTER 6

1. 4
2. -2
3. 2,-5
4. One
5. Two
6. One
7. $-x^3 + x^2 + x$
8. $-2x^2 + 4xy + 7y^2$
   $7y^2 + 4xy - 2x^2$
9. $r^3s - 3r^2s^2 - 3rs^3 + s^4$
10. $12a^2$
11. $3a$
12. 0
13. $4x^2 - 2x$
14. $2x^2 + 5x + 3$
15. $2x^3 + 6x^2 + x + 1$
16. $4y^4 + 6y^3 - 9y^2 - 17$
17. $2x$
18. $+4x^2$
19. 0
20. $5a + 10$
21. $x^2 - 2x - 1$
22. $-2x^2 + x - 5$
23. $-2x^2 + 7x - 17$
24. $-3x$
25. $3x^2 - 4x + 5$
26. $2x + 3$
27. x^5
28. 2^{11}
29. $a^4b^6c^3$
30. $-10a^2$
31. $12x^5$
32. $-12a^3b^4c^4$
33. $-8x + 12$
34. $2a^3 - 6a^2 + 12a$
35. $+3x^4 - 9x^3 + 12x^2$
36. $2x^3 - 12x^2 + 16x$
37. $5z^3 - 17z^2 + 31z - 10$
38. $a^3 - 6a^2 + 12a - 8$
39. y^7
40. xy^2
41. $\frac{x}{z}$
42. $\frac{c^3}{b^3}$
43. $4a^2$
44. $-2b$
45. $-25xz^2$
46. $x + 3$
47. $-x^2 + 3y + 4$
48. $-3xy + 2$
49. $2x - 3$
50. $3a - 4b$

CHAPTER 7
EXERCISE 7-2 (Page 245)

1. $1 - 2x^2$
2. $x + 2y$
3. $x + y$
4. $x + 1$
5. $x^2 + 2x - 5$
6. $x^2 + y^2$
7. $x - 2y$
8. $3x + 4$
9. $x^2 - 5x + 6$
10. $3b^2 - b + 7$
11. $z + x - y$
12. $x^2 - 2x - 1$
13. 3
14. 6
15. $7x$
16. $3x^2$
17. $2x$
18. $3x^2$
19. $5xy^2$
20. $4x^2y$
21. $5xyz$
22. a^2b
23. xyz
24. $7x^2$
25. $7(x + y)$
26. $4(x - y)$
27. $10a$
28. $21x$
29. $3a(b - 2c)$
30. $10ab$
31. $6x^2$
32. $6x(x + 1)$
33. $6(x^2 - 2xy + y^2)$
34. $a(3x^2 - 10x - 7)$
35. $3b^2(1 - ab - b^2)$
36. $4a(3a^2 + 4a + 2)$
37. $(x + 3)(a + b)$
38. $(x - 3)(y - z)$
39. $(x + 2)(x + 3)$
40. $(x - 3)(x - 3)$

EXERCISE 7-3 (Page 250)

1. $x^2 + 8x + 15$
2. $y^2 + 5y + 4$
3. $x^2 - 5x - 14$
4. $y^2 + 4y - 12$
5. $z^2 - 5z - 14$
6. $z^2 + 5z - 24$
7. $x^2 - 5x + 4$
8. $x^2 - 9x + 14$
9. $x^2 - 1$
10. $y^2 - 4$
11. $2x^2 + 11x + 5$
12. $5x^2 - 17x - 12$
13. $3x^2 + 10x - 8$
14. $3x^2 - 17x - 6$
15. $4x^2 - 1$
16. $6x^2 - 13x + 6$
17. $15x^2 + 4x - 35$
18. $6a^2 - ab - 2b^2$
19. $x^4 + 4x^2 + 3$
20. $6a^4 - a^2b^2 - b^4$
21. $18 + 3x - x^2$
22. $6x^2 + 23x + 7$
23. $x^2 + 12x + 36$
24. $4x^2 + 12x + 9$
25. $y^2 - 14y + 49$
26. $4y^2 - 20y + 25$
27. $x^4 + 12x^2 + 36$
28. $4a^4 - 12a^2 + 9$
29. $4a^2 - 12ab + 9b^2$
30. $9a^4 - 6a^2b^2 + b^4$

EXERCISE 7-4 (Page 257)

1. $(x + 2)(x + 1)$
2. $(y + 3)(y + 1)$
3. $(x + 2)(x + 3)$
4. $(y + 5)(y + 1)$
5. $(z + 2)(z + 2)$
6. $(z + 1)(z + 4)$
7. $(x + 1)(x + 9)$
8. $(y + 3)(y + 3)$
9. $(x + 3)(x + 6)$
10. $(y + 5)(y + 7)$
11. $(x - 3)(x - 3)$
12. $(y - 1)(y - 9)$
13. $(x - 2)(x - 5)$
14. $(z - 1)(z - 10)$
15. $(x - 1)(x - 7)$
16. $(y - 2)(y - 3)$
17. $(x - 2)(x - 4)$
18. $(y - 1)(y - 3)$
19. $(y - 1)(y - 6)$
20. $(z - 4)(z - 9)$
21. $(y + 2)(y - 1)$
22. $(z + 6)(z - 3)$
23. $(x + 8)(x - 3)$
24. $(x + 6)(x - 2)$
25. $(x + 6)(x - 3)$
26. $(x + 7)(x - 3)$
27. $(x + 4)(x - 2)$
28. $(x + 12)(x - 5)$
29. $(y + 3)(y - 2)$
30. $(y - 3)(y + 2)$
31. $(y - 5)(y + 2)$
32. $(x - 4)(x + 2)$
33. $(x - 9)(x + 2)$
34. $(y - 4)(y + 1)$
35. $(x - 4)(x + 2)$
36. $(x - 7)(x + 1)$
37. $(z - 8)(z + 3)$
38. $(y - 9)(y + 3)$
39. $(x - 5)(x + 3)$
40. $(z - 16)(z + 3)$
41. $(y + 8)(y - 8)$
42. $(z + 5)(z - 5)$
43. $(c + 10)(c - 10)$
44. $(x + 9)(x - 9)$
45. $(3 + x)(3 - x)$
46. $(4 + y)(4 - y)$
47. $(2x + 5)(2x - 5)$
48. $(5x + y)(5x - y)$
49. $(11 + x)(11 - x)$
50. $(x + y)(x - y)$
51. $(6x + 5y)(6x - 5y)$
52. $(2y + 3)(2y - 3)$
53. $(x_3 + 1)(x_3 - 1)$
54. $(2z + 1)(2z - 1)$
55. $(1 + x)(1 - x)$
56. $(x^3 + 1)(x^3 - 1)$
57. $(x + \frac{1}{2})(x - \frac{1}{2})$
58. $(x + \frac{1}{3})(x - \frac{1}{3})$

EXERCISE 7-5 (Page 264)

1. $(2x + 1)(x + 2)$
2. $(2y + 3)(y + 2)$
3. $(2x + 3)(x + 1)$
4. $(3x + 1)(x + 3)$
5. $(2x + 3)(x + 4)$
6. $(3x + 4)(x + 2)$
7. $(6x + 1)(x + 4)$
8. $(2x + 1)(2x + 1)$
9. $(2x - 3)(2x - 1)$
10. $(7x - 1)(x - 2)$
11. $(4x - 1)(2x - 3)$
12. $(3x - 2)(x - 1)$
13. $(2y - 5)(2y - 1)$
14. $(5x - 2)(2x - 1)$
15. $(3y - 7)(y - 2)$
16. $(x - 3)(9x - 4)$
17. $(2x - 3)(x + 5)$
18. $(2x + 7)(2x - 1)$
19. $(3y - 4)(y + 3)$
20. $(4y - 1)(y + 5)$
21. $(3x - 5)(x + 2)$
22. $(2x + 1)(3x - 2)$
23. $(3x - 4)(2x + 1)$
24. $(3x - 4)(5x + 7)$
25. $(2x + 3)(x - 5)$
26. $(3x + 1)(x - 2)$
27. $(3x + 4)(x - 3)$
28. $(2x - 3)(9x + 2)$
29. $(5x + 1)(3x - 1)$
30. $(3x - 4)(2x + 1)$
31. $(3x - 4)(5x + 2)$
32. $(3x - 4)(6x + 5)$
33. $(4x + 5)(4x - 5)$
34. $(2x + 1)(2x - 1)$
35. $(5x + 7)(5x - 7)$
36. $(3x + 2)(3x - 2)$

EXERCISE 7-6 (Page 267)

1. $3(x + 3)(x - 3)$
2. $8x(x + 1)(x - 1)$
3. $(z^2 + 4)(z + 2)(z - 2)$
4. $2(x + 2y)(x - 2y)$
5. $a(x + y)(x - y)$
6. $3(x + 3y)(x - 3y)$
7. $z(z + 1)(z - 1)$
8. $(x^2 + 1)(x + 1)(x - 1)$
9. $4(a + 3)(a - 3)$
10. $4(x + 2)(x - 2)$
11. $x(x + 5)(x - 5)$
12. $(ab + 11)(ab - 11)$
13. $2(x - 2)(x - 2)$
14. $3(y + 3)(y + 3)$
15. $a(x + 2)(x + 1)$
16. $3a(y - 2)(y - 1)$
17. $2y(2x - 3)(x - 1)$
18. $4(x - 4)(x + 3)$
19. $x(x + 5)(x + 2)$
20. $2a(y + 3)(y - 2)$
21. $(x^2 + 2)(x + 1)(x - 1)$
22. $(x + 1)(x - 1)(x + 3)(x - 3)$
23. $xy(x + 5y)(x + 5y)$
24. $x^2(x - 6)(x - 6)$

EXERCISE 7-7 (Page 271)

1. $x = 0$; $x = 7$
2. $y = 0$; $y = -4$
3. $x = 0$; $x = 4$
4. $y = 0$; $y = -3$
5. $x = 0$; $x = \frac{3}{2}$
6. $y = 0$; $y = -\frac{2}{3}$
7. $x = 3$; $x = -3$
8. $x = 4$; $x = -4$
9. $z = 3$; $z = -3$
10. $x = \frac{5}{3}$; $x = -\frac{5}{3}$
11. $x = \frac{3}{2}$; $x = -\frac{3}{2}$
12. $x = \frac{1}{6}$; $x = -\frac{1}{6}$
13. $x = 1$; $x = 3$
14. $x = -1$; $x = -2$
15. $x = 2$; $x = 5$
16. $x = 7$; $x = -2$
17. $x = 1$, $x = -6$
18. $y = -4$; $y = 6$
19. $y = 1$; $y = 2$
20. $x = 7$; $x = -3$
21. $z = 2$; $z = 9$
22. $y = -3$; $y = 5$
23. $y = -1$; $y = 7$
24. $x = 7$; $x = -5$
25. $x = -\frac{1}{2}$; $x = -2$
26. $x = \frac{1}{3}$; $x = -3$
27. $x = -\frac{4}{3}$; $x = 1$
28. $x = -\frac{3}{2}$; $x = 1$
29. $x = -\frac{1}{3}$; $x = -3$
30. $x = -\frac{1}{5}$; $x = -2$
31. $y = -\frac{3}{2}$; $y = -2$
32. $x = -3$; $x = -3$
33. $x = \frac{3}{4}$; $x = -4$
34. $x = 1$; $x = 1$
35. $y = \frac{1}{2}$, $y = \frac{1}{2}$
36. $x = -\frac{2}{3}$, $x = \frac{4}{5}$
37. $x = \frac{5}{2}$, $x = -\frac{4}{3}$
38. $y = -5$, $y = 3$
39. $x = -\frac{1}{2}$, $x = 10$
40. $y = -4$, $y = 3$

REVIEW EXERCISES (Page 272)
CHAPTER 7

1. $(6x^3)$
2. $(x^2 - 4)$
3. $(x + 2)$
4. $(x - 3)$
5. 5
6. $3x^2$
7. $2xy$
8. 5
9. $3x(x + 2)$
10. $ab(b^2 + a)$
11. $6(-2x^2 + 5x + 1)$
12. $4x(2x - 3 - 6x^2)$
13. $(x - 4)(x + 3)$
14. $x^2 - 4x - 21$
15. $y^2 + 7y - 18$
16. $2x^2 - 5x - 12$
17. $6y^2 - y - 15$
18. $x^2 - 4$
19. $9x^2 - 16$
20. $a^2 - b^2$
21. $x^2 - 9$
22. $y^2 + 10y + 25$
23. $z^2 - 6z + 9$

REVIEW EXERCISES
CHAPTER 7 (Cont'd)

24. $4x^2 - 12x + 9$
25. $(x + 5)(x - 3)$
26. $(x - 7)(x + 1)$
27. $(x + 2)(x + 6)$
28. $(y - 7)(y - 5)$
29. $(2x + 1)(x - 4)$
30. $(3x - 1)(2x + 3)$
31. $(2x - 1)(2x - 1)$
32. $(2x - 3)(3x + 2)$
33. $(x + 5)(x - 5)$
34. $(2y + 5)(2y - 5)$
35. $(4 + 3x)(4 - 3x)$
36. $3(x + 1)(x - 1)$
37. $z^3(z + 1)(z - 1)$
38. $(y^2 + 4)(y + 2)(y - 2)$
39. $3(x - 4)(x + 3)$
40. $2x(x - 1)(x - 1)$
41. $y = 0, y = 7$
42. $z = 0, z = 4$
43. $y = 4, y = -4$
44. $x = 5, x = -5$
45. $x = 5, x = -3$
46. $x = -7, x = 1$
47. $x = 6, x = 2$
48. $y = -7, y = -5$
49. $x = \frac{1}{2}, x = -4$
50. $x = \frac{3}{2}, x = -\frac{1}{3}$

CHAPTER 8
EXERCISE 8-1 (Page 279)

1. $\dfrac{(3)(x)}{(4)(y)}$
2. $\dfrac{2(x + 2)}{4(x + 3)}$
3. $\dfrac{x(x + 2)}{2(x + 2)}$
4. $\dfrac{2x(x + 3)}{3(x - 3)}$
5. $\dfrac{(x + 3)(x - 3)}{(x + 3)(x + 3)}$
6. $\dfrac{(x - 3)(x + 2)}{(x + 2)(x + 3)}$
7. $\dfrac{(x + 3)(x - 2)}{(x - 2)(x - 2)}$
8. $\dfrac{2(x - 4)(x - 3)}{2(x + 5)(x - 4)}$
9. $\dfrac{2(x + 2)(x - 1)}{4(x + 1)(x - 1)}$
10. $\dfrac{x(x - 1)}{3(x + 1)(x - 1)}$
11. $+\dfrac{3x}{4}$
12. $-\dfrac{2x}{7}$
13. $-\dfrac{x - 2}{-2 + x}$
14. $\dfrac{-x(-1 + x)}{(x + 1)(x - 1)}$
15. $+\dfrac{2(-1 + y)}{3(y - 1)}$
16. $+\dfrac{2(x + 1)}{3(x + 1)}$
17. $+\dfrac{+2x(-1 + x)}{+3(x - 1)}$

EXERCISE 8-2 (Page 284)

1. $\dfrac{x}{y}$
2. $\dfrac{1}{3x}$
3. $\dfrac{x}{4}$
4. 2
5. $\dfrac{1}{2}$
6. $\dfrac{b}{4}$
7. $\dfrac{x}{4y^2}$
8. $\dfrac{-2}{3xy}$
9. $-\dfrac{2x}{3y}$
10. $\dfrac{5x}{-6y^2}$
11. $\dfrac{3}{4}$
12. $2x$
13. $\dfrac{1}{2x}$
14. $-\dfrac{1}{2}$
15. $\dfrac{3}{5}$
16. $\dfrac{x + 1}{x - 1}$

EXERCISE 8-2 (Cont'd)

17. $\dfrac{1}{4}$
18. $\dfrac{x+y}{2}$
19. $\dfrac{2(x-y)}{3}$
20. $\dfrac{(x+y)}{2}$
21. $\dfrac{2}{x-2}$
22. $\dfrac{(x+2)}{(x-2)}$
23. $\dfrac{1}{x+1}$
24. $\dfrac{x-2}{x+2}$
25. $\dfrac{7}{x+1}$
26. $\dfrac{2}{y-1}$
27. $\dfrac{x-2}{x+2}$
28. $\dfrac{x-2}{x+2}$
29. $\dfrac{4}{x+3}$
30. 3
31. $3(x-1)$
32. $\dfrac{x}{x-3}$
33. $\dfrac{x-1}{x+4}$
34. $\dfrac{x+3}{x-3}$
35. $\dfrac{y-5}{y+2}$
36. $\dfrac{x+2}{x-3}$
37. $-\dfrac{2}{3}$
38. -1
39. $\dfrac{x+5}{x-8}$
40. $\dfrac{x-1}{3x+2}$

EXERCISE 8-3 (Page 290)

1. $\dfrac{3}{5}$
2. $\dfrac{6x^2}{5}$
3. $\dfrac{5}{3a}$
4. $6x$
5. $\dfrac{3}{xy}$
6. x^2
7. $\dfrac{x}{6}$
8. $\dfrac{a^2}{6}$
9. $\dfrac{2x^3}{y^4}$
10. $2r$
11. $\dfrac{2}{3}$
12. $\dfrac{x-1}{x+3}$
13. $\dfrac{3}{2}$
14. 1
15. $\dfrac{3}{4}$
16. $\dfrac{1}{3}$
17. $\dfrac{15x}{2(x-2)(x-3)}$
18. $\dfrac{2}{x-2y}$
19. 2
20. $\dfrac{x-4}{8x}$
21. $\dfrac{1+3x}{2(x-2)}$
22. $\dfrac{3(x+1)(x+2)}{x+3}$
23. $\dfrac{x(x+1)}{x+4}$
24. $\dfrac{3x}{x-3}$
25. 1

EXERCISE 8-4 (Page 293)

1. $\dfrac{1}{x}$
2. $\dfrac{1}{3x}$
3. x
4. $\dfrac{4b}{5a}$
5. 12
6. $15x$
7. $6x$
8. $\dfrac{y^2}{3}$
9. $\dfrac{y}{21x}$
10. $\dfrac{xy}{8z}$
11. $\dfrac{x}{14}$
12. $2x$
13. $\dfrac{3}{2}$
14. $\dfrac{2}{3}$
15. $\dfrac{2(x-y)}{3}$
16. $\dfrac{3}{x-1}$
17. $\dfrac{2}{x+y}$
18. $\dfrac{5}{6}$
19. 1
20. $\dfrac{1}{2}$
21. $3(x+3)$
22. $\dfrac{x+1}{x-1}$

EXERCISE 8-5 (Page 298)

1. x
2. $\frac{5}{x}$
3. $\frac{5x}{12}$
4. $\frac{3}{5x}$
5. $\frac{5}{x+1}$
6. $\frac{2}{x-1}$
7. 1
8. 2
9. $x+y$
10. $x-4$
11. $\frac{6x+1}{3}$
12. $x+1$
13. 9
14. $\frac{1}{x-1}$
15. $\frac{2x+1}{x+3}$
16. $\frac{1}{x-4}$

EXERCISE 8-6 (Page 311)

1. $6x$
2. $-25x$
3. $2x$
4. $2(x-1)$ or $2x-2$
5. $6x$
6. $x(x-2)$ or x^2-2x
7. $4(x+1)$ or $4x+4$
8. $(x-1)(x+1)$ or x^2-1
9. $(x-2)(x-2)$ or x^2-4x+4
10. $2(x-2)(x-3)$ or $2x^2-10x+12$
11. $12x$
12. $12xy$
13. x^2
14. $4x^2$
15. $50x^2$
16. $4(x+1)$
17. $x(x+1)$
18. $3x(x-1)$
19. $(x+y)(x-y)$
20. $3(x+5)(x-5)$
21. $(x-2)(x-3)$
22. $(x-3)(x-2)(x+3)$
23. $\frac{19x}{30}$
24. $\frac{5}{4x}$
25. $\frac{19}{10x}$
26. $\frac{15x+8}{25x^2}$
27. $\frac{3y^2-2x^2}{18x^2y^2}$
28. $\frac{3x^3+5x-15}{10x^2}$
29. $\frac{-3(2x-1)}{4}$
30. $\frac{17}{6(3x-y)}$
31. $\frac{-y-16}{3(y+4)(y-4)}$
32. $\frac{4x(2x-1)}{(x+2)(x-2)}$
33. $\frac{x}{3(y+2)}$
34. $\frac{13}{3x(x+1)}$
35. $\frac{24}{(x+6)(x-6)}$
36. $\frac{5}{3(2x-1)}$
37. $\frac{3x-2}{(x-2)(x-1)}$
38. $\frac{7x-13}{(x-3)(x+2)(x-1)}$
39. $\frac{-x^2+3x+2}{(x+6)(x+1)}$
40. $\frac{2(-x+1)}{(x+3)(x-2)}$
41. $\frac{9x^2+7x+5}{(x+3)(x-1)(x+2)}$
42. $\frac{2x}{(x+1)(x-2)}$
43. $\frac{x^2+3}{x}$
44. $\frac{3t+2}{t+2}$
45. $\frac{x^2+x+1}{x+1}$
46. $\frac{x^2-x-1}{x-1}$
47. $\frac{4(-x+3)}{x-2}$
48. $\frac{x^2}{x-1}$
49. $\frac{3x+8}{x+1}$
50. $\frac{cx^2-2cx+c-1}{c}$

EXERCISE 8-7 (Page 316)

1. $\dfrac{3}{(x+1)(x-1)}$
2. $\dfrac{9x}{2(9x-2)}$
3. $\dfrac{3+5x}{3-5x}$
4. $\dfrac{x^2+1}{(x+1)(x-1)}$
5. $\dfrac{(x+2)(x-2)(x+2)(x-2)}{4x^3}$
6. $\dfrac{(x-2)(x-2)}{3(x+3)(x+3)}$
7. $\dfrac{x+1}{x-1}$
8. $\dfrac{x-2y}{2x+y}$

EXERCISE 8-8 (Page 322)

1. $x = 18$
2. $x = \dfrac{45}{26}$
3. $x = 60$
4. $x = -96$
5. $x = 38\dfrac{2}{3}$
6. $x = 24$
7. $x = 14$
8. $x = 3$
9. $x = \dfrac{7}{5}$
10. $x = -5$
11. $x = 2$
12. $x = 19$
13. $x = 10$
14. $x = 4$
15. $x = 4$
16. $y = 4$
17. $x = 2$
18. $x = 6$
19. $y = -\dfrac{1}{3}$
20. $x = \dfrac{1}{4}$
21. $x = 4$
22. $x = 26$
23. $x = 6$
24. $x = \dfrac{5}{4}$
25. $\dfrac{15}{25}$
26. $N = 18$
27. $N = 3$
28. $37\dfrac{1}{2}, 62\dfrac{1}{2}$
29. $x = -20$
30. $N = 73$

EXERCISE 8-9 (Page 328)

1. 3 hours
2. 5.5 hours
3. 7.2 hours (or 7 hours 12 minutes)
4. 24 minutes
5. 1 hour
6. 8 minutes
7. 12 hours
8. 12 hours
9. $10\dfrac{2}{3}$ hours (or 10 hours 40 minutes)
10. $2\dfrac{3}{4}$ days
11. (a) 40 minutes (b) No
12. 6 minutes
13. 21.5 minutes
14. $13\dfrac{1}{3}$ minutes (or 13 minutes 20 seconds)
15. Kathy: $245 LuAnn: $325 Marilyn: $65

REVIEW EXERCISES (Page 329)
CHAPTER 8

1. $24x^2y^3$
2. $3x^2 - 9x$
3. 2
4. $x^2 - 1$
5. $\dfrac{1}{x+1}$
6. $\dfrac{x+1}{x+2}$
7. $\dfrac{x+1}{x-3}$
8. $\dfrac{3x^2}{2y}$
9. $\dfrac{x-1}{2(x-3)}$
10. $\dfrac{x^2}{4}$
11. $\dfrac{x}{2}$
12. 2
13. 1
14. $\dfrac{(x+9)}{(x+5)(x-5)}$
15. $\dfrac{-2}{x(x-1)}$
16. $\dfrac{3x^2 - 26x + 11}{(x+2)(x+1)(x-3)(x-1)}$
17. $\dfrac{2x-1}{x-1}$
18. $\dfrac{(3x+1)(x+2)}{x(2x+1)}$
19. $x = -1$
20. $x = -\dfrac{1}{4}$
21. $x = 5$
22. $x = 6$

CHAPTER 9
EXERCISE 9-1 (Page 334)

1. $2xy$
2. $3x^2$
3. $(x + 1)$
4. $(2x - 1)$
5. $4y$
6. $-3x$
7. $2x^2$
8. $-x^3$
9. $12x^3y^2$
10. $5xy^2z^3$
11. 4.58
12. 2.65
13. 11.83
14. 1.41
15. 8.94
16. 6.32
17. -10.05
18. -11.53
19. 11.00
20. 8.19
21. 2.4
22. 3.2
23. 3.3
24. 3.9
25. 4.4
26. 5.1
27. 7.1
28. 7.5
29. 10.5
30. 11.4

EXERCISE 9-2 (Page 341)

1. $2\sqrt{2}$
2. $3\sqrt{2}$
3. $6\sqrt{2}$
4. $2\sqrt{3}$
5. $2\sqrt{5}$
6. $3\sqrt{5}$
7. $4x\sqrt{2}$
8. $8x^2\sqrt{x}$
9. $5x^2y^2\sqrt{2xy}$
10. $3xy\sqrt{3y}$
11. $7xy^2z^3\sqrt{2y}$
12. $7x^2y^3z^4\sqrt{xyz}$
13. $6\sqrt{6}$
14. $16x\sqrt{3}$
15. $10x\sqrt{5x}$
16. $2x\sqrt{2}$
17. $2x\sqrt{2x}$
18. $xy^2\sqrt{7x}$
19. $-6x\sqrt{x}$
20. $-3x^2\sqrt{6}$
21. $-4y\sqrt{5x}$
22. $-xy\sqrt{5y}$
23. $-xz^2\sqrt{yz}$
24. $-2x^3y^4\sqrt{x}$
25. $\frac{3}{4}$
26. $\frac{2}{3}$
27. $\frac{3}{5}$
28. $\frac{5}{6}$
29. $\frac{1}{4}$
30. $\frac{1}{2}$
31. 2
32. $\frac{\sqrt{2}}{2}$
33. $\frac{\sqrt{5}}{2}$
34. $\frac{20\sqrt{2}}{7}$
35. $\frac{10\sqrt{2}}{7}$
36. $\frac{7\sqrt{7}}{27}$
37. $\frac{3x}{2}$
38. $\frac{18x}{5}$
39. $\frac{2\sqrt{30}}{9}$
40. $\frac{14x^2\sqrt{10}}{10y}$ or $\frac{7x^2\sqrt{10}}{5y}$
41. $2x^2\sqrt{15}$
42. $\frac{y^2}{2x}$
43. $\frac{x\sqrt{2}}{3}$
44. $\frac{x}{y}$
45. $\frac{1}{3}$
46. $\frac{3\sqrt{3y}}{5x}$
47. $\frac{x\sqrt{30}}{9}$
48. $\frac{\sqrt{19}}{20y^2}$
49. $\frac{x^2yz\sqrt{xz}}{5}$
50. $\frac{\sqrt{xyz}}{10}$

EXERCISE 9-3 (Page 345)

1. $5\sqrt{3}$
2. $-5\sqrt{5}$
3. $3\sqrt{2}$
4. 0
5. $-2\sqrt{5}$
6. $3\sqrt{2}$
7. $4\sqrt{5} + 3\sqrt{2}$
8. $5\sqrt{6} + 3\sqrt{3}$
9. $5\sqrt{3} - \sqrt{5}$
10. $2\sqrt{5}$
11. $3\sqrt{2}$
12. $-2\sqrt{3}$
13. $\sqrt{5}$
14. $3\sqrt{6}$
15. $3\sqrt{2}$
16. 0
17. $\sqrt{6}$
18. $2\sqrt{5}$
19. $3\sqrt{2}$
20. 0
21. $6\sqrt{2}$
22. $4\sqrt{2}$
23. $4\sqrt{2}$
24. $5\sqrt{6}$

EXERCISE 9-3 (Cont'd)

25. $-5\sqrt{3}$ 26. $3\sqrt{2}$ 27. $-9\sqrt{2}$ 28. $\sqrt{2}$
29. $3\sqrt{5}$ 30. $2\sqrt{3}$ 31. $2\sqrt{6}$ 32. $2\sqrt{7}$
33. 0 34. $7\sqrt{5}$ 35. $\sqrt{11}$ 36. $-7\sqrt{6}$
37. $\sqrt{3}$
38. $\dfrac{14\sqrt{2}}{3}$ 39. $10 - 6\sqrt{3}$ 40. $-2\sqrt{5}$ 41. $3\sqrt{5}$
42. $2\sqrt{6} - \dfrac{9}{4}$ 43. $5\sqrt{10}$ 44. $\dfrac{\sqrt{3}}{2} + \dfrac{4\sqrt{7}}{5} - 2\sqrt{3}$
45. $110 - \dfrac{\sqrt{35}}{4}$ 46. $3\sqrt{2} - 2\sqrt{6} + \dfrac{10\sqrt{5}}{7}$

EXERCISE 9-4 (Page 350)

1. 3 2. 21 3. x 4. $x + 1$
5. 4 6. 6 7. 6 8. 10
9. $6\sqrt{5}$ 10. $14\sqrt{10}$ 11. $-12\sqrt{3}$ 12. $+4\sqrt{5}$
13. $8\sqrt{15}$ 14. $2\sqrt{6}$ 15. 24 16. 120
17. $18\sqrt{10}$ 18. $12\sqrt{21}$ 19. $-2\sqrt{6}$ 20. $-6\sqrt{10}$
21. $3\sqrt{2}$ 22. $2\sqrt{3}$ 23. $8\sqrt{3}$ 24. $10\sqrt{3}$
25. $18\sqrt{5}$ 26. $4\sqrt{5}$ 27. $-12\sqrt{2}$ 28. $5\sqrt{2}$
29. $8\sqrt{5}$ 30. $21\sqrt{5}$ 31. $\sqrt{6}$ 32. $18\sqrt{2}$
33. $30\sqrt{7}$ 34. $27\sqrt{7}$ 35. $\sqrt{10}$ 36. $8\sqrt{11}$
37. 2 38. 4 39. $3\sqrt{2}$ 40. $2\sqrt{3}$
41. $\dfrac{\sqrt{3}}{2}$ 42. $\dfrac{2\sqrt{2}}{5}$ 43. 1 44. $\sqrt{2}$
45. $\dfrac{\sqrt{21}}{6}$ 46. $\dfrac{3\sqrt{6}}{5}$ 47. $6x$ 48. $4x\sqrt{3}$
49. $2x\sqrt{5y}$ 50. $6xy^2\sqrt{2xy}$ 51. $42x^3y^2$ 52. $20x^2y\sqrt{3xy}$
53. 3 54. 20 55. 18 56. $4x$
57. $8x$ 58. $18x$ 59. $2\sqrt{6} - 2\sqrt{15}$ 60. $\sqrt{6} + \sqrt{10}$
61. 30 62. 15 63. $10\sqrt{2} - 20\sqrt{3}$ 64. $24 + 4\sqrt{6}$

EXERCISE 9-5 (Page 359)

1. 3 2. 16 3. 2 4. $\dfrac{1}{2}$
5. $\dfrac{5}{2}$ 6. 14 7. 2 8. 3
9. $2\sqrt{3}$ 10. $3\sqrt{2}$ 11. $\sqrt{3}$ 12. $3\sqrt{2}$
13. $5\sqrt{2}$ 14. $9\sqrt{2}$ 15. $x\sqrt{5}$ 16. 3
17. 2 18. $x\sqrt{3}$ 19. $2y\sqrt{x}$ 20. $3y\sqrt{x}$
21. $\dfrac{\sqrt{3}}{3}$ 22. $\dfrac{\sqrt{2}}{2}$ 23. $\dfrac{\sqrt{6}}{3}$ 24. $\sqrt{21}$

EXERCISE 9-5 (Cont'd)

25. $\sqrt{3}$
26. $\dfrac{2\sqrt{5}}{5}$
27. $\dfrac{2\sqrt{6}}{3}$
28. $\dfrac{\sqrt{10}}{4}$
29. $\dfrac{1}{2}$
30. $\dfrac{\sqrt{2}}{2}$
31. $\dfrac{\sqrt{2}}{2}$
32. $\dfrac{\sqrt{3}}{3}$
33. $\dfrac{\sqrt{5}}{5}$
34. $2\sqrt{2}$
35. $\dfrac{5\sqrt{2}}{2}$
36. $3\sqrt{3}$
37. $\sqrt{3}$
38. $2\sqrt{3}$
39. $\dfrac{2\sqrt{2}}{3}$
40. $\dfrac{\sqrt{2}}{3}$
41. $\dfrac{\sqrt{7}}{4}$
42. $\sqrt{2}$
43. $\dfrac{\sqrt{15}}{6}$
44. $\dfrac{2\sqrt{5}}{3}$
45. $2\sqrt{5}$
46. $\dfrac{\sqrt{6}}{3}$
47. $\sqrt{3}$
48. $4\sqrt{2}$
49. $4\sqrt{3}$
50. $-\sqrt{5}$
51. $\sqrt{5}$
52. $\sqrt{6}$
53. $4\sqrt{10}$
54. $\dfrac{5\sqrt{3}}{6}$
55. $3\sqrt{10}$
56. $-\sqrt{6}$

EXERCISE 9-6 (Page 365)

1. x = 49
2. x = 3
3. x = 0
4. x = 12
5. x = 6
6. No Solution
7. No Solution
8. a = 10
9. x = 48
10. 32
11. No Solution
12. x = 5
13. x = 3
14. x = 10
15. x = 3
16. x = 2
17. x = 5
18. y = $\dfrac{49}{16}$
19. x = 3
20. x = 4
21. x = 3
22. x = 5
23. x = 85
24. x = 16
25. x = 7
26. x = 2
27. x = 4

REVIEW EXERCISES (Page 366)
CHAPTER 9

1. 5.4
2. 11.40
3. 4.8
4. 8.5
5. 14
6. $3x\sqrt{6}$
7. $4x^2y^3\sqrt{7xy}$
8. $\dfrac{\sqrt{21}}{7}$
9. $\dfrac{3\sqrt{2}}{2}$
10. $\dfrac{4\sqrt{10}}{5}$
11. $2\sqrt{5}$
12. $13\sqrt{2}$
13. $7\sqrt{3}$
14. $\sqrt{2}$
15. $-\sqrt{6}$
16. $18\sqrt{2}$
17. $84\sqrt{3}$
18. $\dfrac{\sqrt{6}}{6}$
19. 20
20. 2
21. $\dfrac{1}{2}$
22. $2\sqrt{6}$
23. $\sqrt{3}$
24. x = 26
25. No Solution
26. N = 64

CHAPTER 10
EXERCISE 10-1 (Page 376)

1. A(5,-1)
2. B(-2,3)
3. C(-3,2)
4. D(1,3)
5. E(0,-4)
6. F(-1,4)
7. G(-3,-2)
8. H(-2,-2)
9. I(-3,0)
10. J(-5,4)

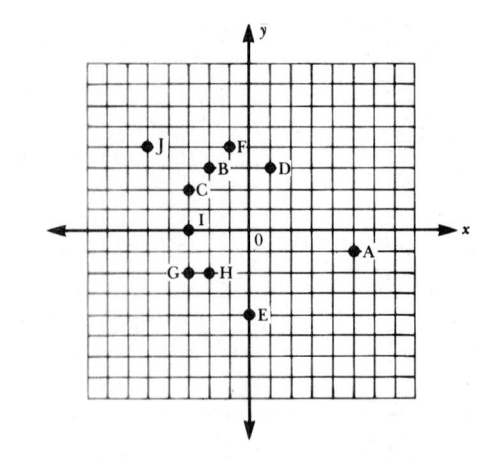

11.

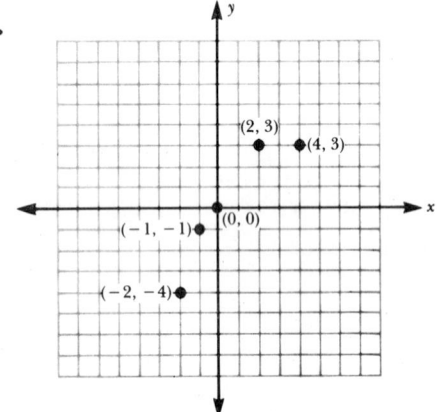

12.

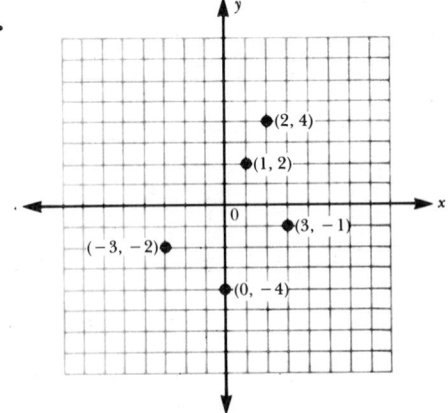

13. (-4,6)
14. (-5,-4)
15. (-9,0)
16. (-12,7)
17. (10,5)
18. (4,-6)
19. (-12,-5)
20. (4,2)
21. (0,-2)
22. (0,8)
23. (11,-3)
24. (8,0)
25. Zero
26. Zero
27 (a) Second Quadrant
27 (b) First Quadrant

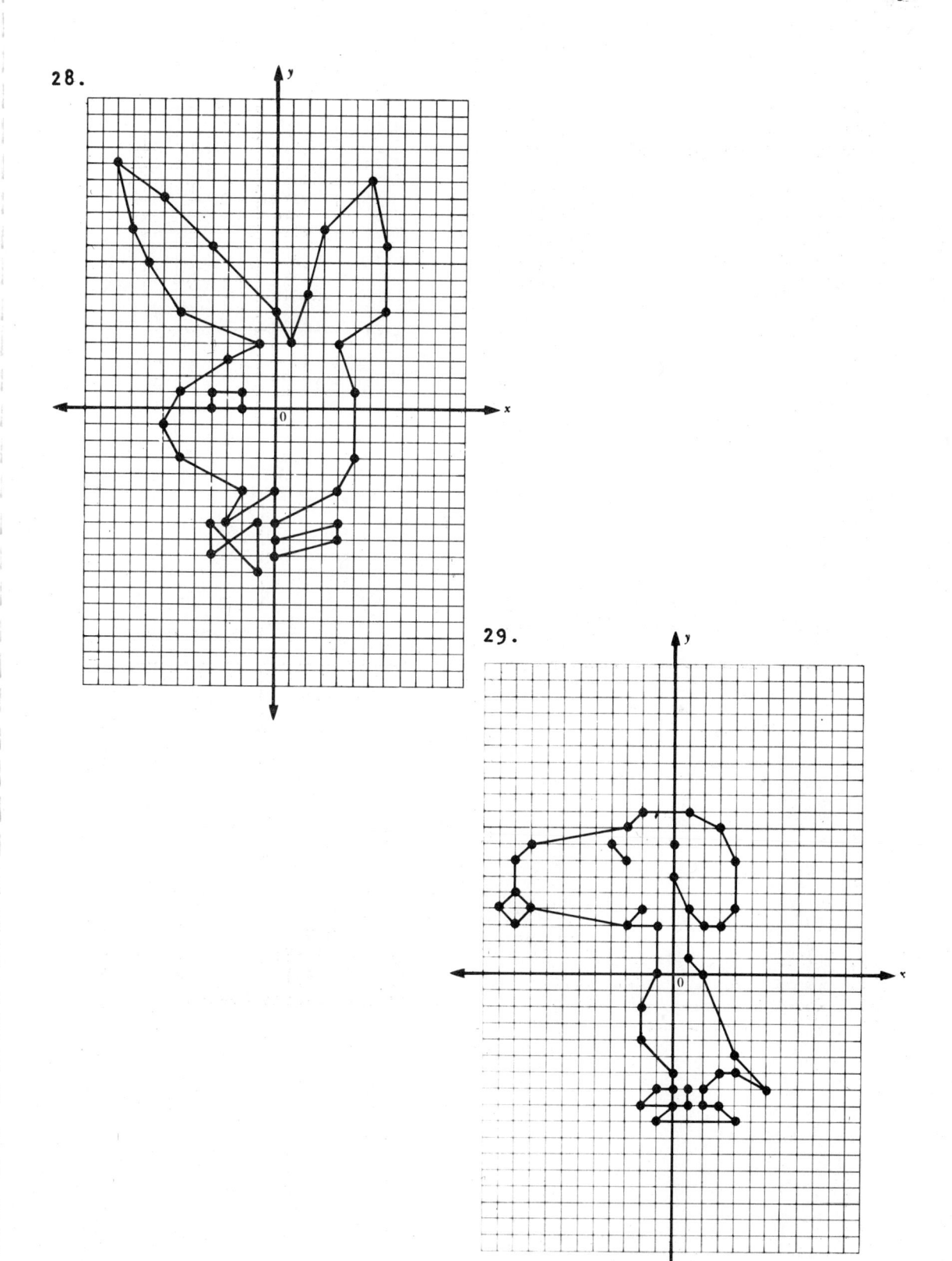

28.

29.

EXERCISE 10-2 (Page 381)

1. Yes
2. No
3. No
4. Yes
5. No
6. No
7. Yes
8. No
9. No
10. Yes
11. y = 7 or (3,7)
12. y = -5 or (2,-5)
13. y = 2 or (4,2)
14. y = -4 or (-3,-4)
15. y = -2 or (1,-2)
16. y = 3 or (7,3)
17. y = -8 or (-8,-8)
18. y = 0 or (0,0)
19. y = -2 or (7,-2)
20. y = 2 or (3,2)
21. y = -1 or (-1,-1)
22. y = 1 or (5,1)
23. y = 3 or (5,3)
24. y = -4 or (12,-4)
25 - 30. These answers will vary.

EXERCISE 10-3 (Page 390)

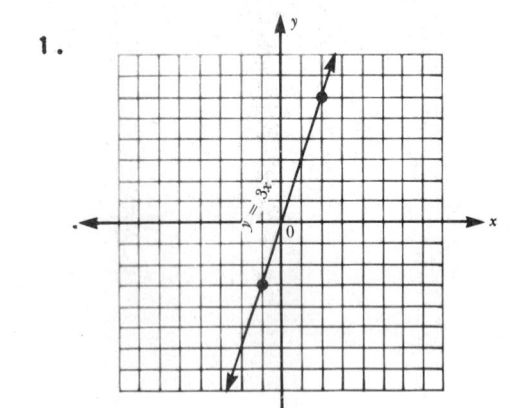

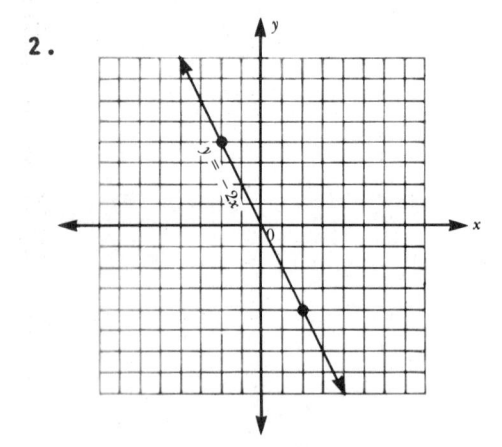

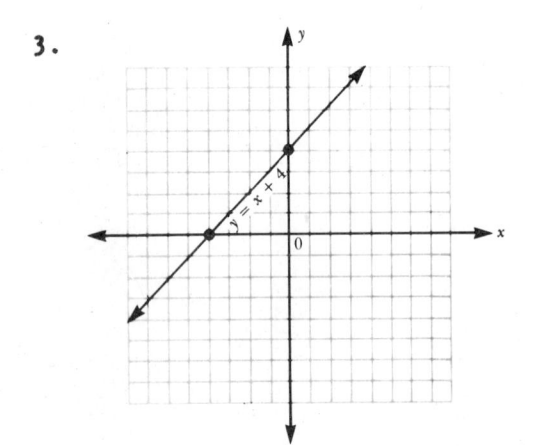

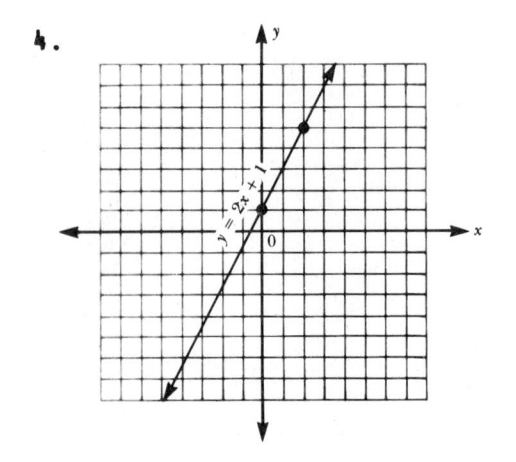

1. [Graph of y = 3x]

2. [Graph of y = -2x]

3. [Graph of y = x + 4]

4. [Graph of y = 2x + 1]

5. $y = -2x - 1$

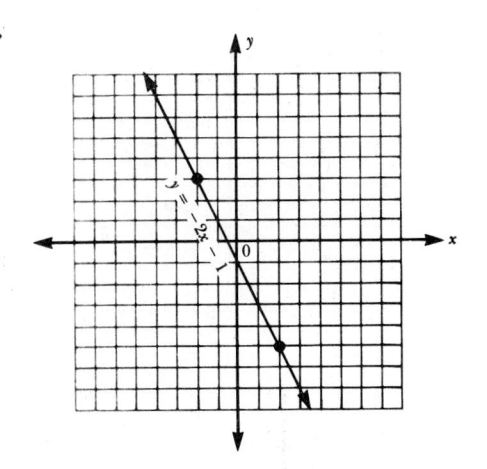

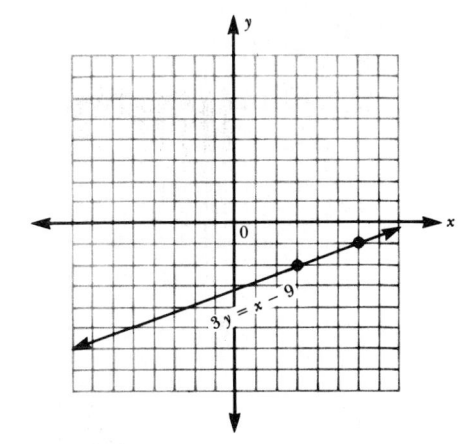

6. $y = 5x - 3$

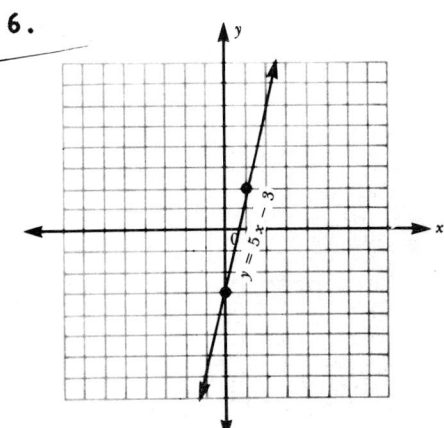

7. $3y = x - 9$

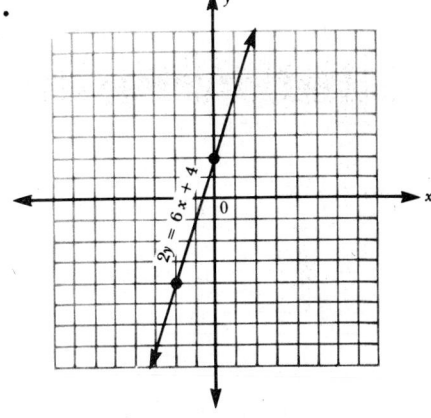

8. $2y = 6x + 4$

9. $y + x = 3$

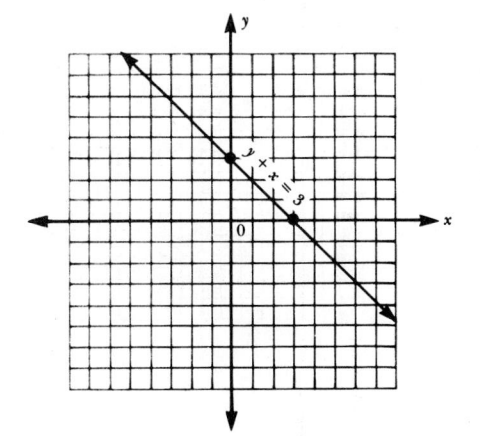

10. $y - x = -3$

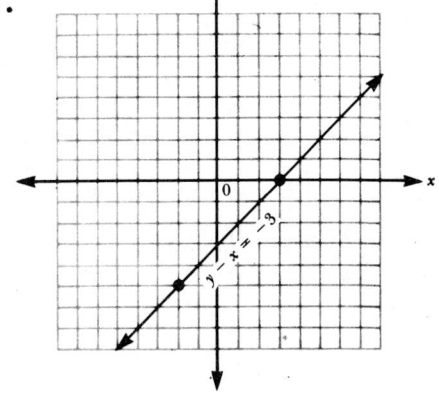

11. 	 $x + 2y = 12$

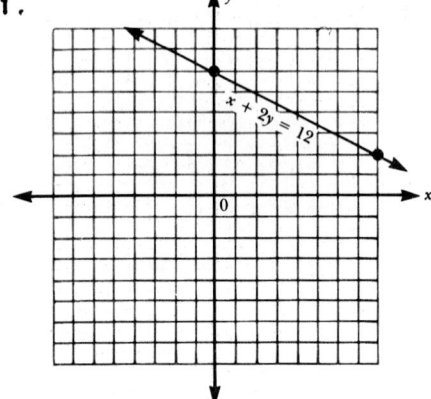

12. 	 $x - 3y = 0$

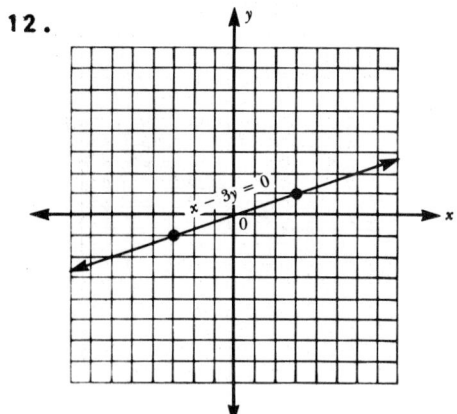

13. 	 $2x + 4y = 12$

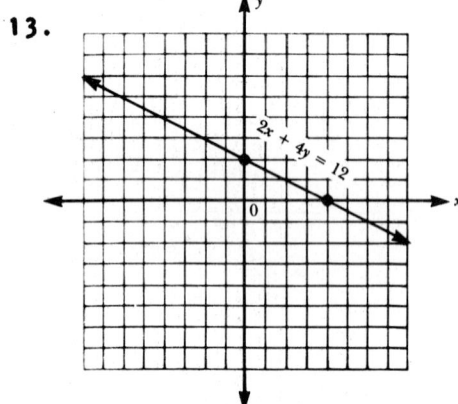

14. 	 $2x + 3y = -6$

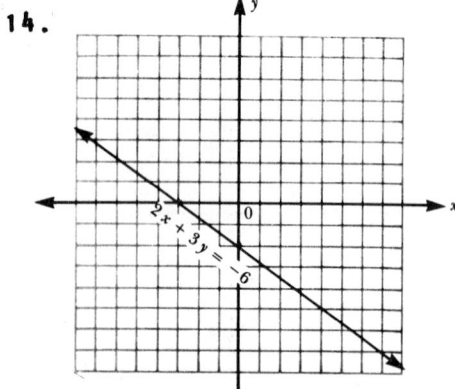

15. 	 $3y - 2x = 12$

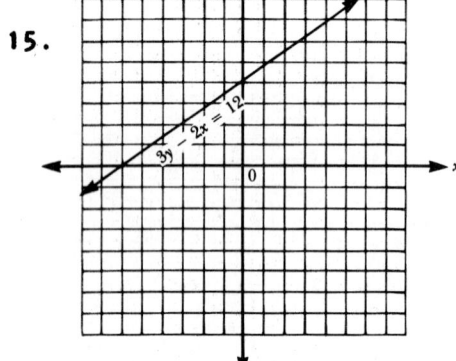

16. 	 $4y - 6x = 24$

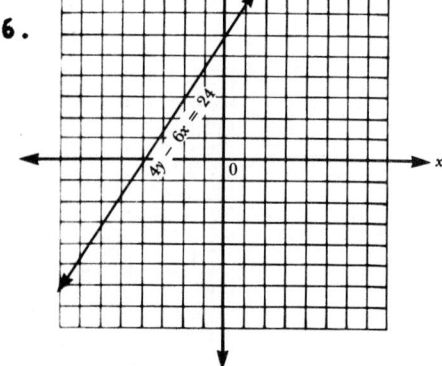

17. $y = \frac{1}{2}x$

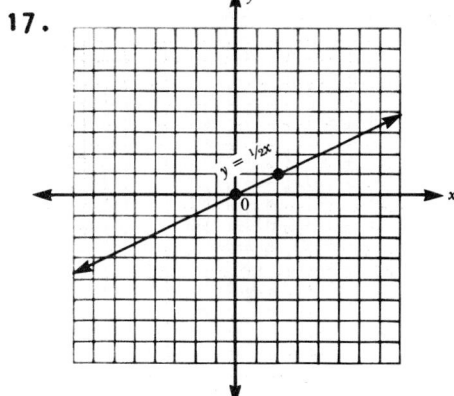

18. $y = \frac{2}{3}x$

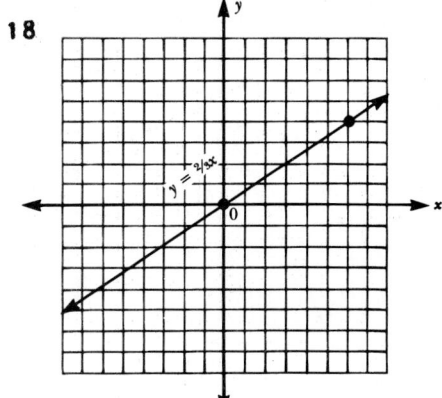

19. $y = -\frac{3}{2}x + 1$

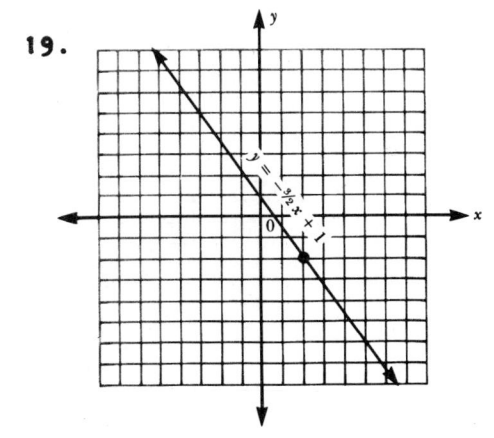

20. $y = -\frac{1}{4}x - 2$

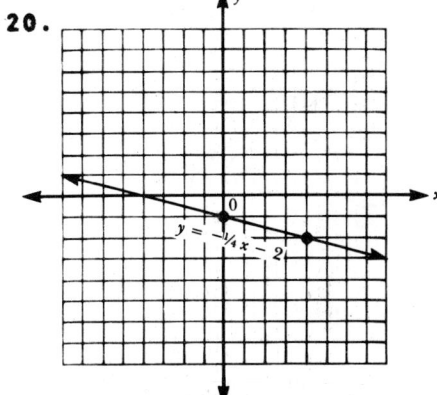

21. $x + y = 3$

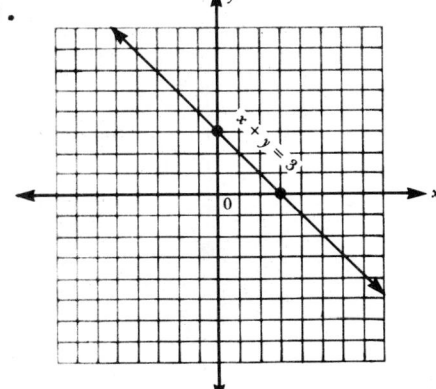

22. $x - y = 5$

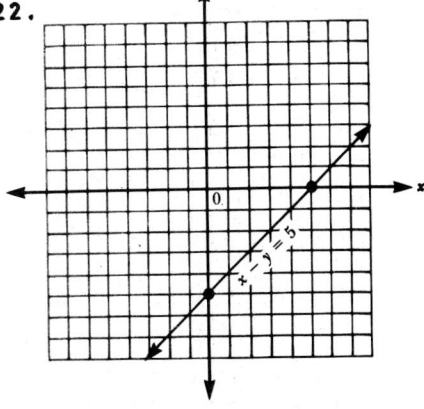

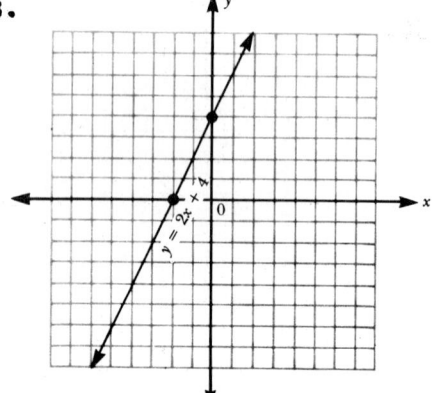

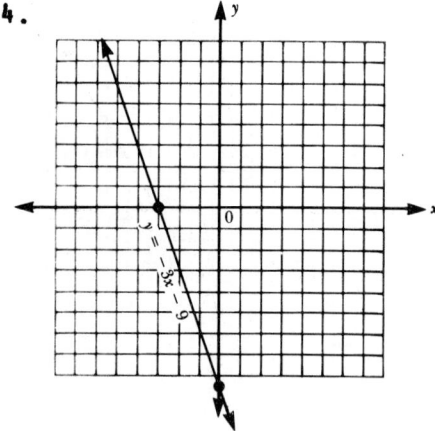

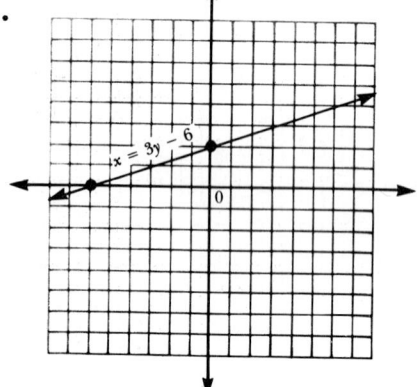

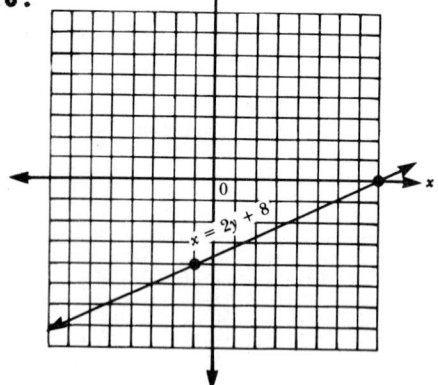

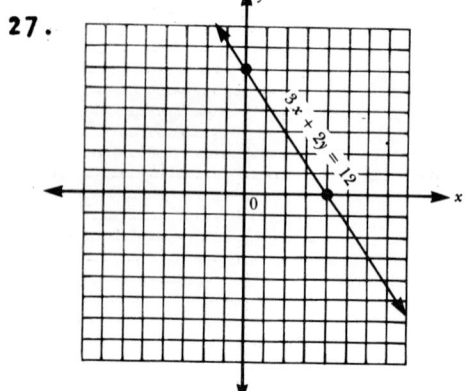

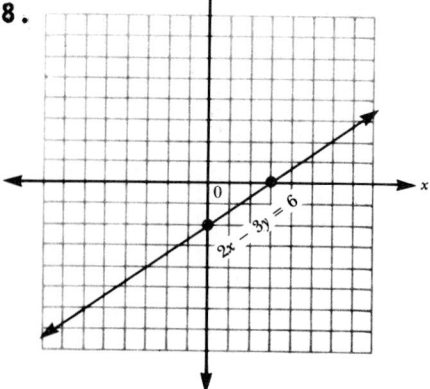

23.

24.

25.

26.

27.

28.

29. $5x - 3y = 15$

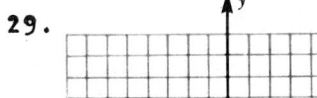

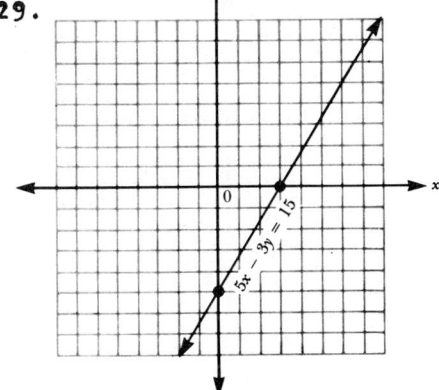

30. $4x + 5y = 10$

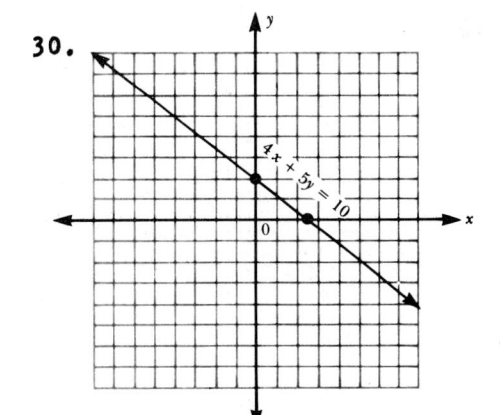

31. $x = 3$

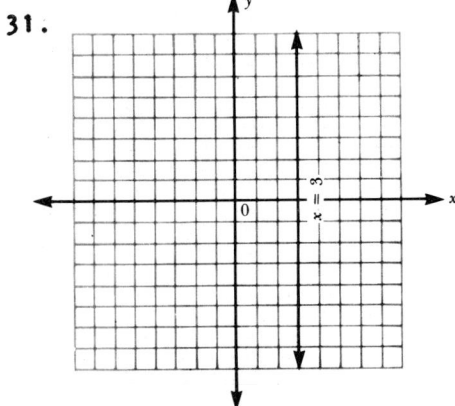

32. $y = -4$

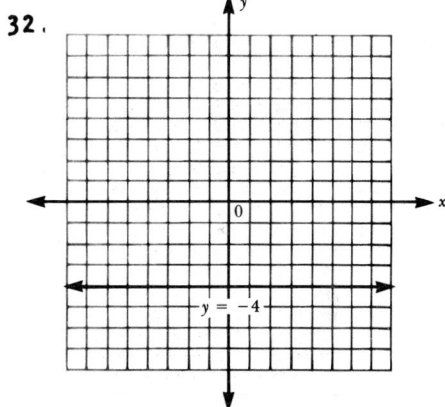

33. $y = 7$

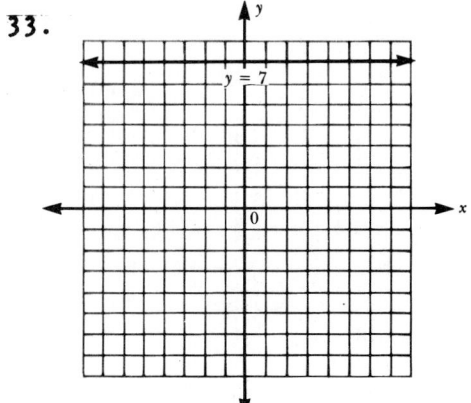

34. $x = -5$

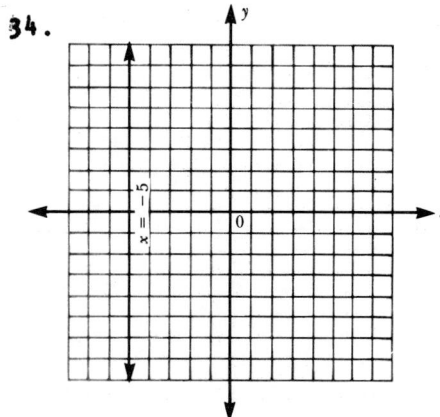

35. $y = 0$

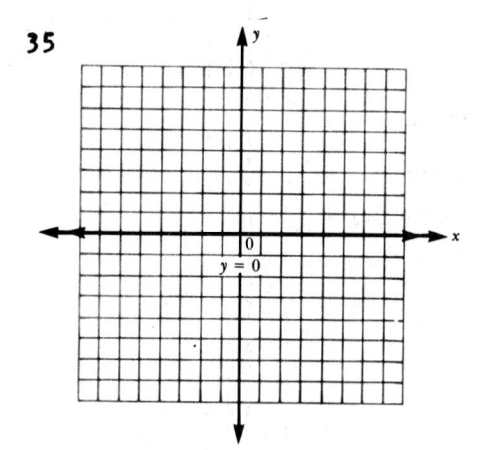

36. $x = 0$

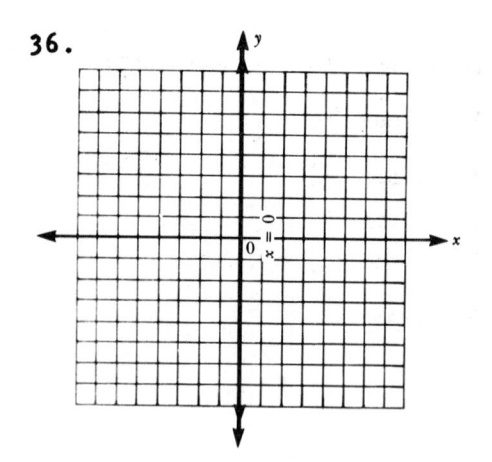

37. $y = 3/5$

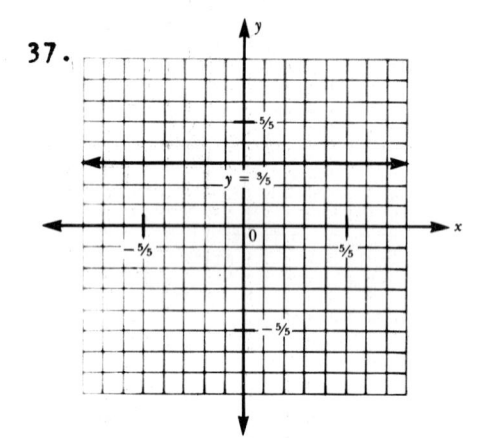

38. $x = 1/2$

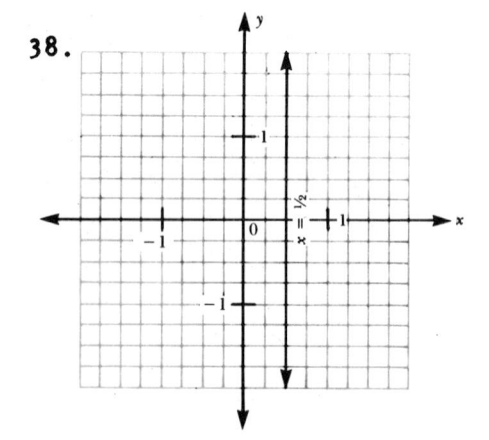

39. $2x - 1 = 0$

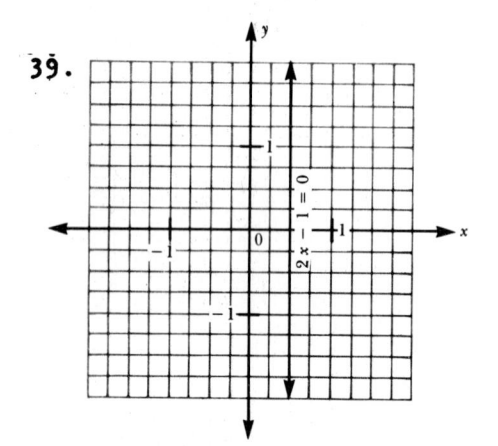

40. $4y = -3$

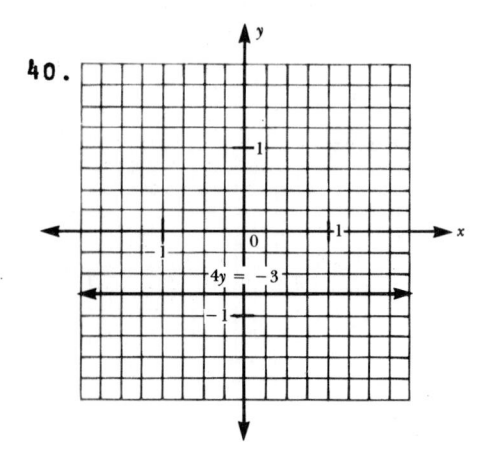

EXERCISE 10-4 (Page 405)

1. Slope (m) = $\frac{1}{2}$
2. Slope (m) = 2
3. Slope (m) = -1
4. Slope (m) = 0
5. Slope (m) = $\frac{4}{-5}$
6. Slope (m) = $\frac{3}{2}$
7. Slope (m) = undefined
8. Slope (m) = -5/3
9. Slope (m) = $\frac{2}{3}$
10. Slope (m) = $\frac{2}{-5}$

11.

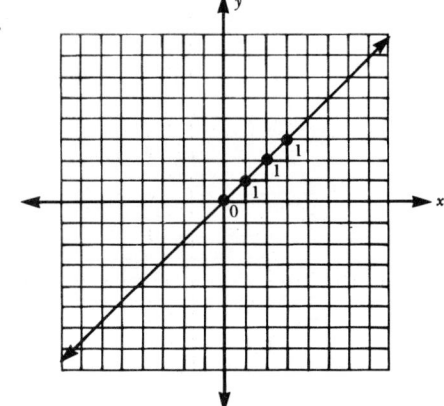

12.

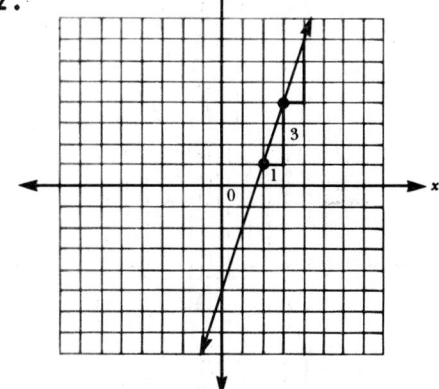

13.

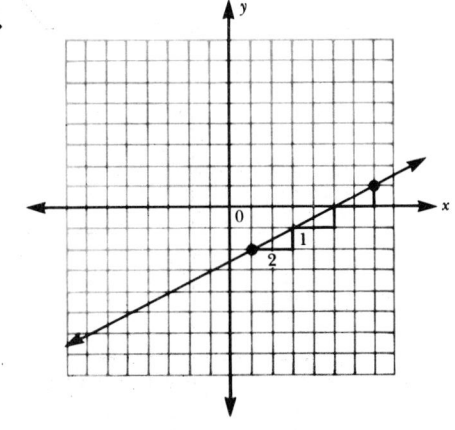

14.

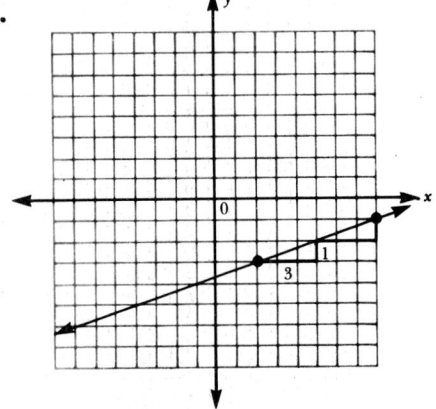

15.
16.
17.
18.
19.
20.

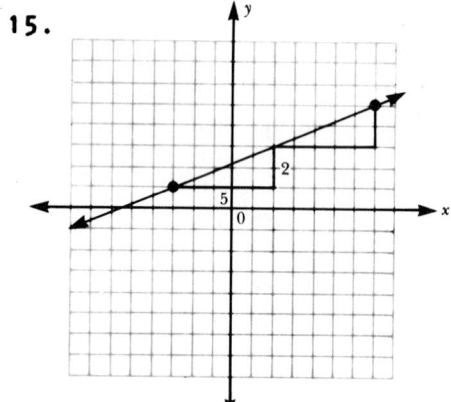

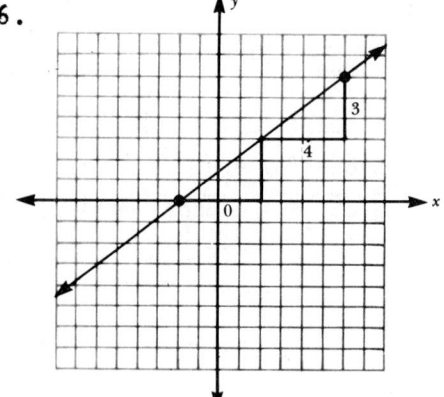

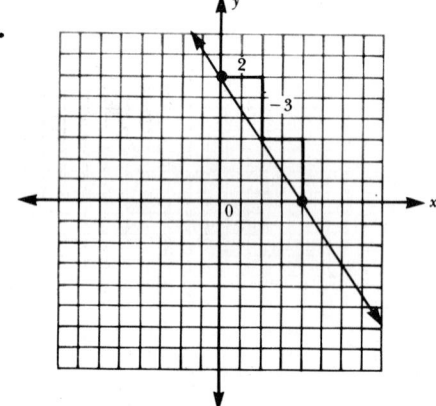

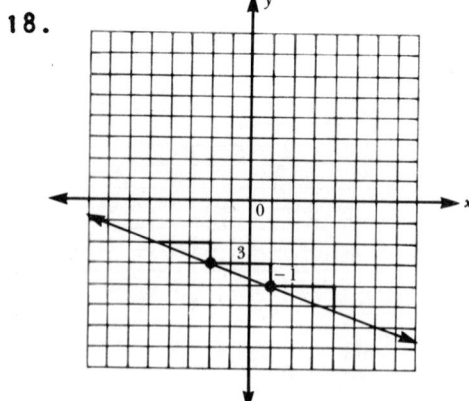

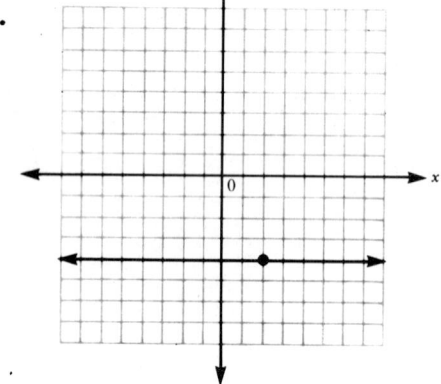

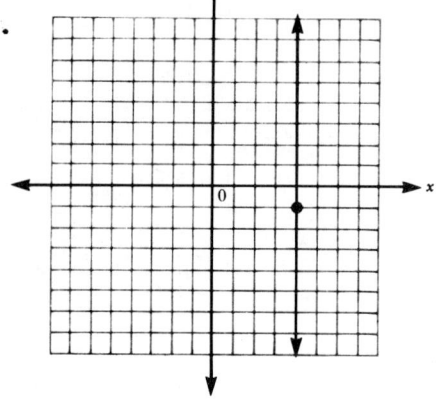

21. $m = 2$, $b = 1$; thus, $(0, 1)$

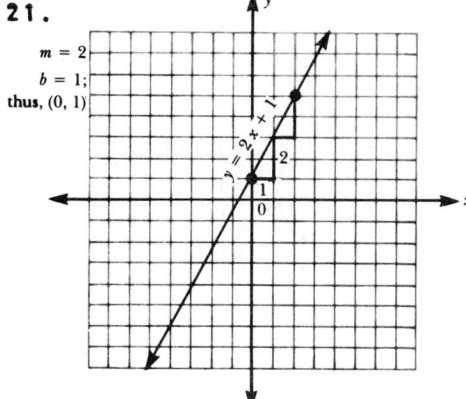

22. $m = 1$, $b = -4$; thus, $(0, -4)$

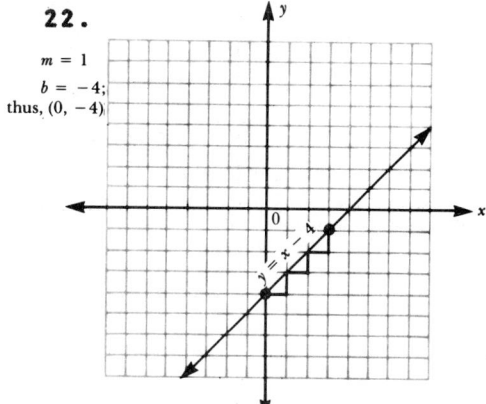

23. $m = 4$, $b = -3$, $(0, -3)$

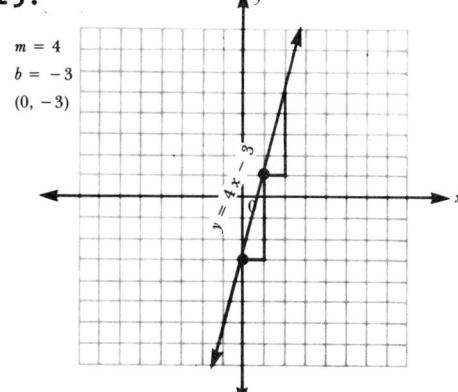

24. $m = 2$, $b = 0$, $(0, 0)$

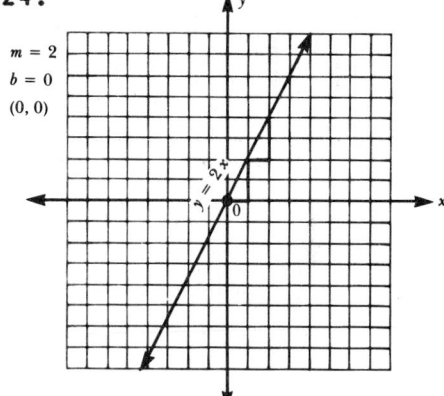

25. $m = 1$, $b = 0$, $(0, 0)$

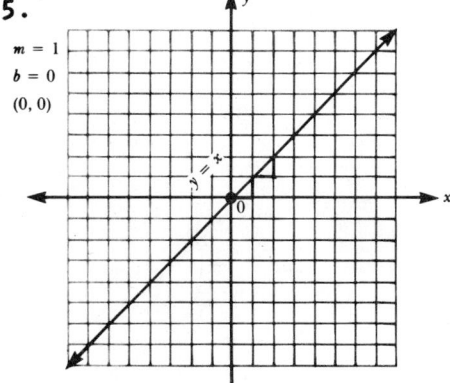

26. $m = -1$, $b = 3$, $(0, 3)$

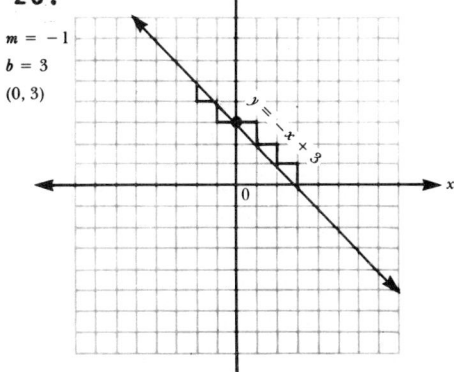

27.
$m = \frac{1}{2}$
$b = 6$
$(0, 6)$

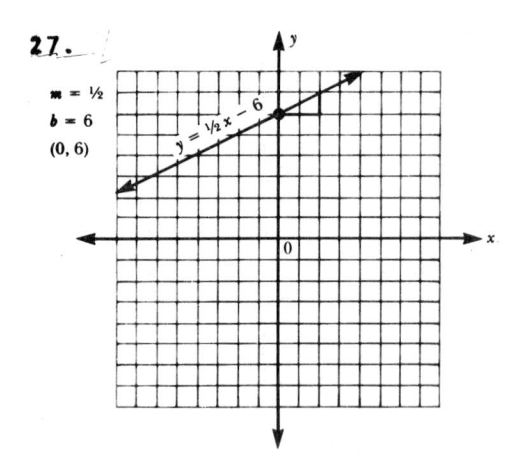

$y = \frac{1}{2}x - 6$

28.
$m = 8$
$b = 0$
$(0, 0)$

$y = 0$

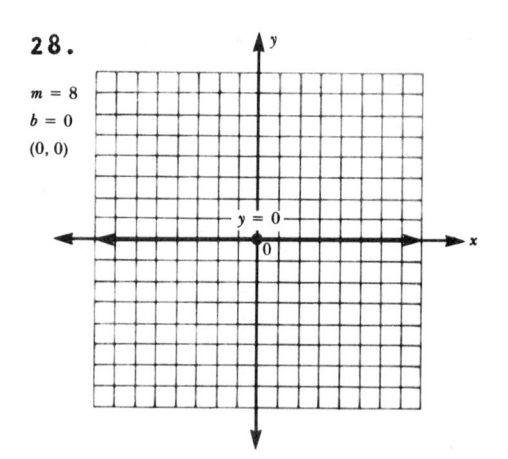

29.
$m = -\frac{2}{3}$
$b = 4$
$(0, 4)$

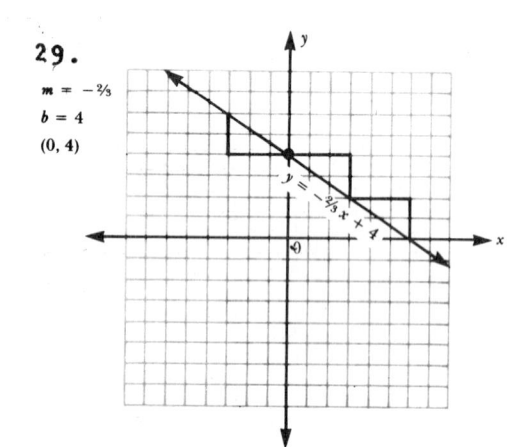

$y = -\frac{2}{3}x + 4$

30.
$m = \frac{3}{5}$
$b = -6$
$(0, -6)$

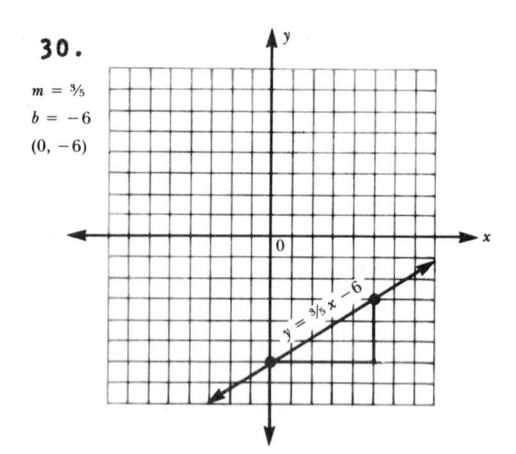

$y = \frac{3}{5}x - 6$

31.
$m = 1$
$b = -3$
$(0, -3)$

$x - y = -3$

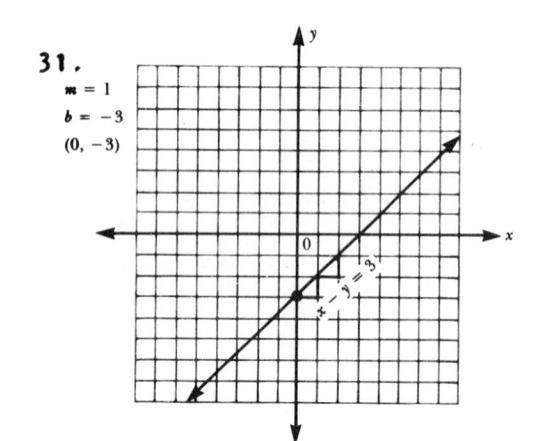

32.
$m = -1$
$b = -3$
$(0, -3)$

$x + y = -3$

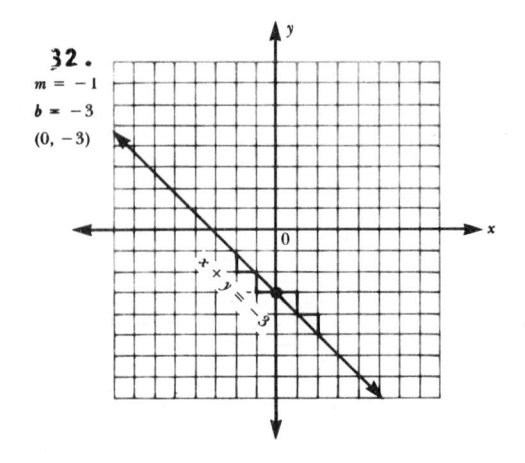

33.
$m = -2/3$
$b = 8/3$
$(0, 8/3)$

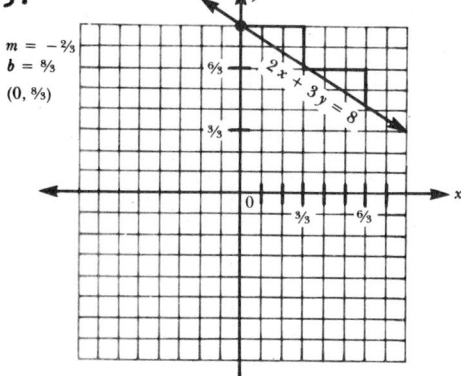

$2x + 3y = 8$

34.
$m = 1/4$
$b = 0$
$(0, 0)$

$x - 4y = 0$

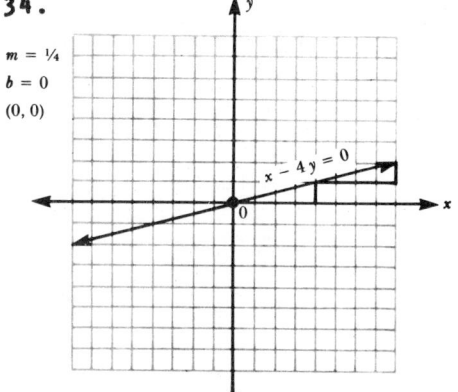

35.
$m = 3/4$
$b = -4$
$(0, -4)$

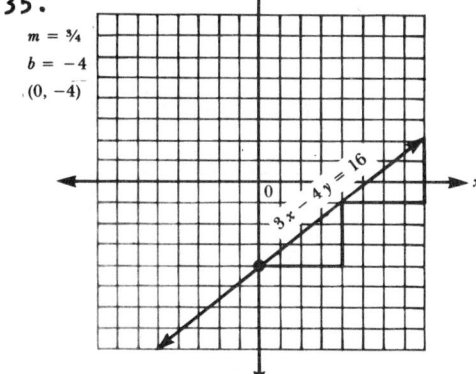

$3x - 4y = 16$

36.
$m = -5/2$
$b = -7/2$
$(0, -7/2)$

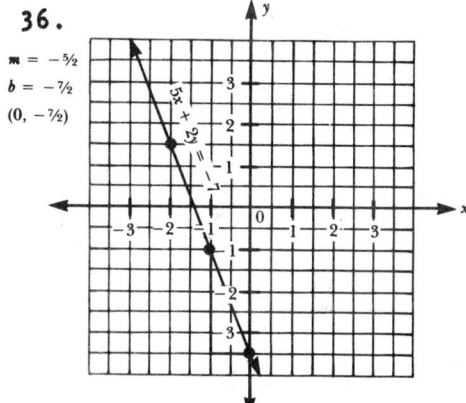

$5x + 2y = -7$

37.
$m = -2/3$
$b = 4/3$
$(0, 4/3)$

$3y = -2x + 4$

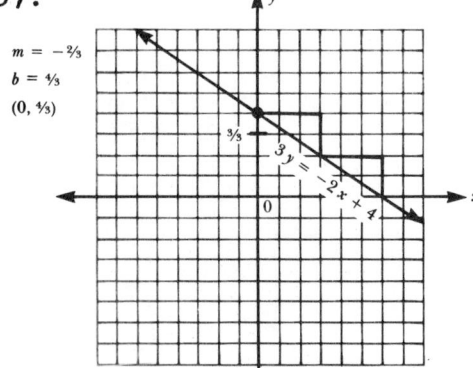

38.
$m = 3/-2$
$b = 4$
$(0, 4)$

$-2y = 3x - 8$

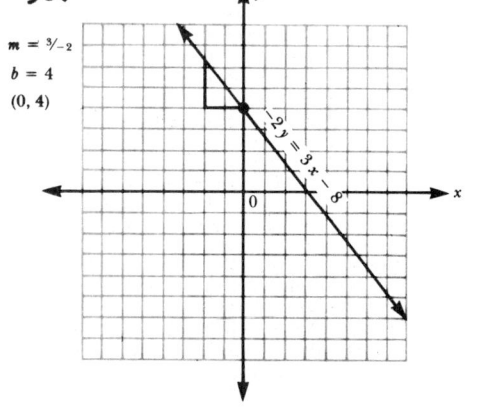

39.
m = 2
b = −6
(0, −6)

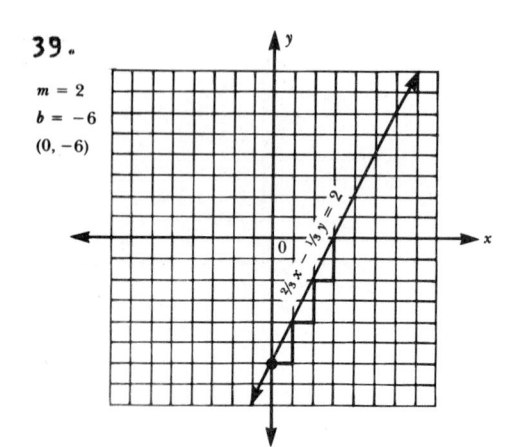

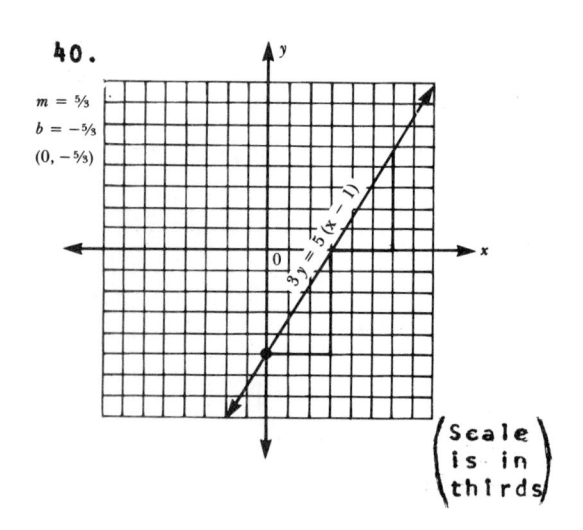

$\frac{2}{3}x - \frac{1}{3}y = 2$

40.
m = 5/3
b = −5/3
(0, −5/3)

$3y = 5(x - 1)$

(Scale is in thirds)

EXERCISE 10-5 (Page 409)

1. $y = 2x - 1$
2. $y = -1$
3. $y = -4x - 7$
4. $y = 3x + 11$
5. $y = \frac{x}{2} - \frac{3}{2}$
6. $y = -\frac{3}{4}x - 2$
7. $y = -\frac{5}{4}x + \frac{3}{2}$
8. $y = -\frac{3}{5}x + \frac{21}{5}$
9. $x = 3$
10. $y = 4$
11. $y = -x + 6$
12. $y = 9x - 19$
13. $y = -\frac{3}{2}x$
14. $y = -x + 2$
15. $y = x + 2$
16. $y = \frac{11}{3}x - 1$
17. $y = \frac{1}{2}x - 2$
18. $y = -\frac{13}{2}x - \frac{29}{2}$
19. $y = \frac{1}{2}x + \frac{1}{2}$
20. $y = -x - 2$
21. $y = 2x + 1$
22. $y = -2x - 3$
23. $y = x - 2$
24. $y = -\frac{2}{3}x$
25. $y = \frac{2}{3}x + 1$

EXERCISE 10-6 (Page 418)

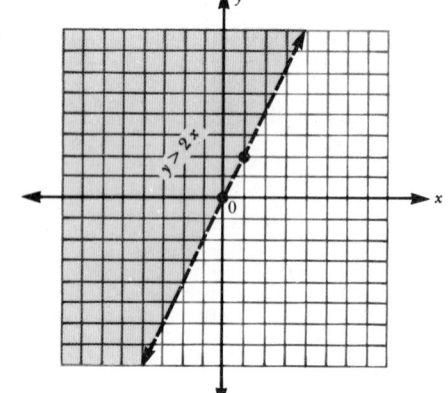

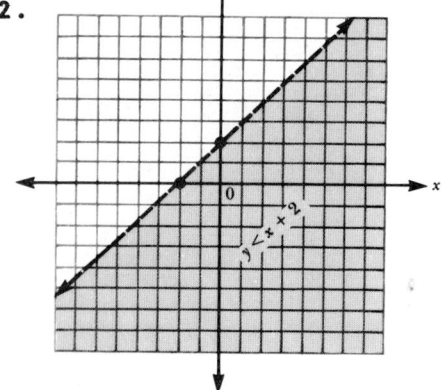

1. $y > 2x$

2. $y < x + 2$

3. [graph: $y \leq 2x - 3$]

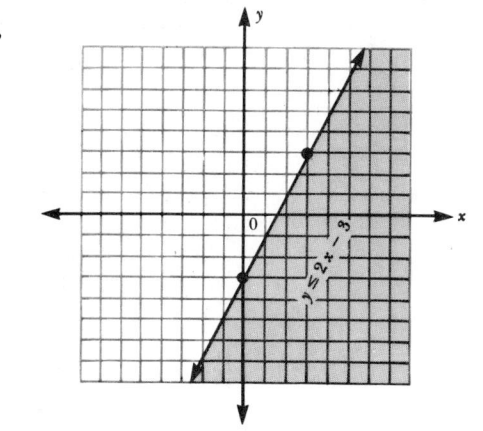

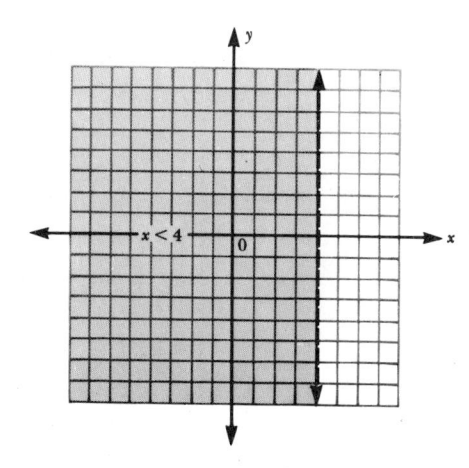

4. [graph: $y \geq 3$]

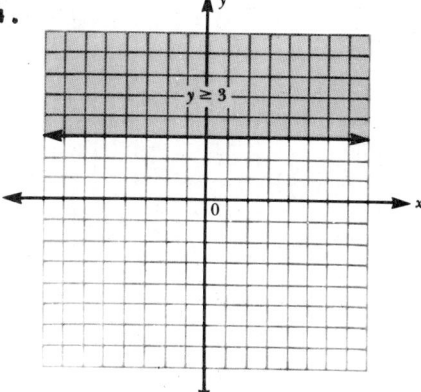

5. [graph: $x < 4$]

6. [graph: $x + y > 3$]

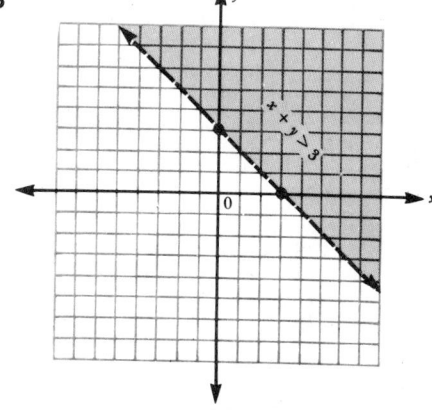

7. [graph: $x - 2y \leq 4$]

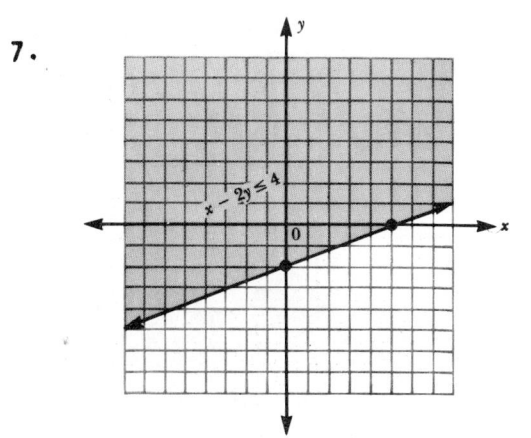

8. [graph: $y - 2x \geq 4$]

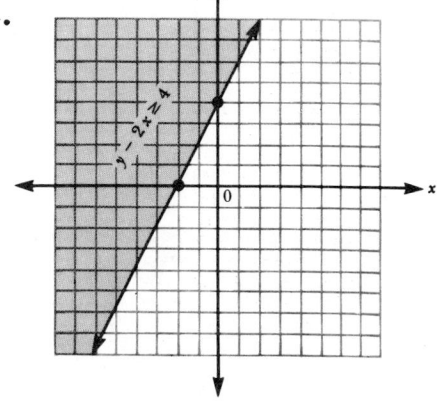

9. $2x + y \geq 0$

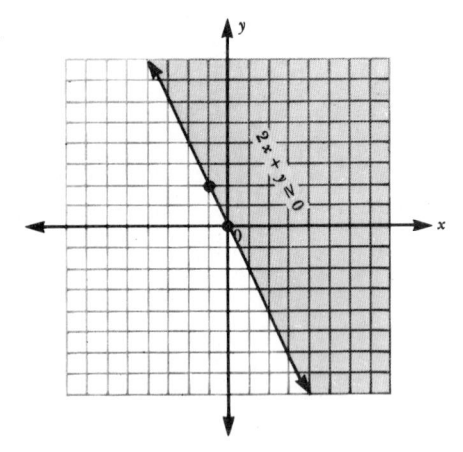

10. $2x + 3y \leq 6$

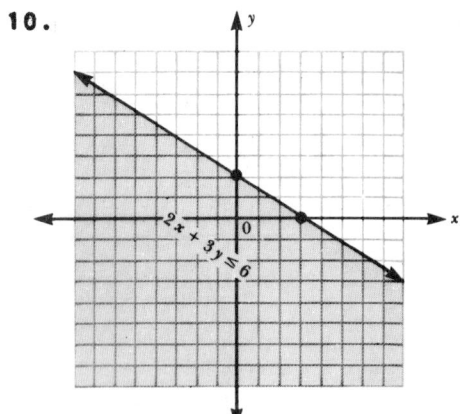

11. $x + 3y - 4 > 0$

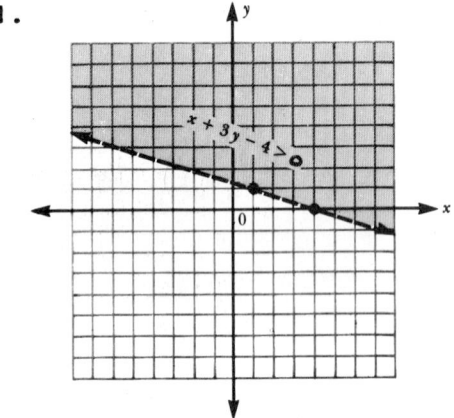

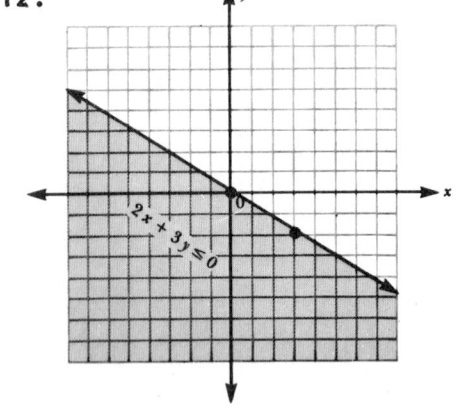

12. $2x + 3y \leq 0$

13. $5 \geq 2x + 3y$

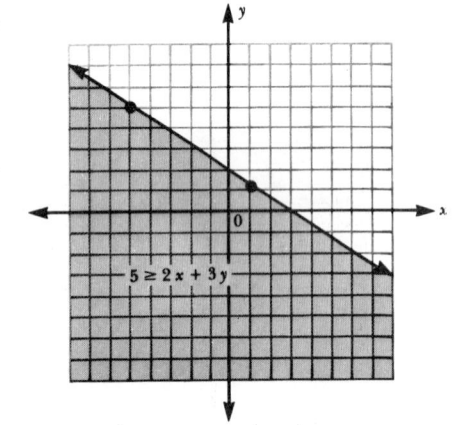

14. $-6 < x - 3y$

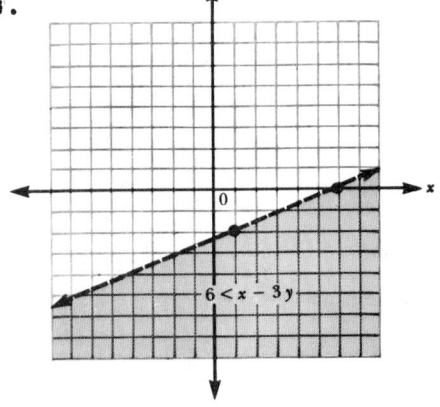

15. $x \leq 2y$

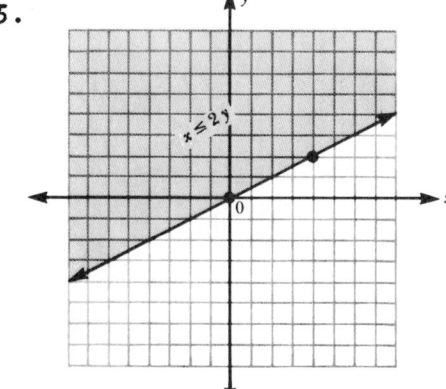

16. $y > 8 + x$

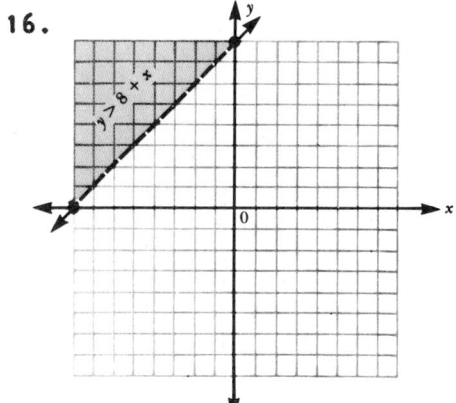

17. $x - y < 2$

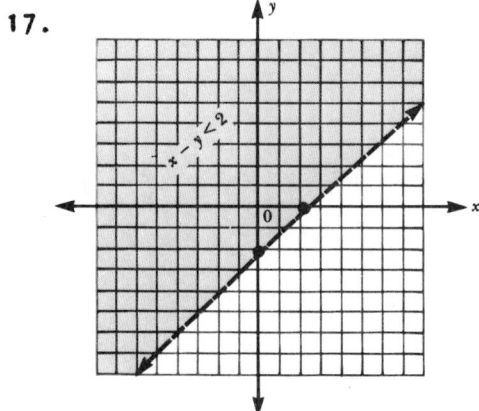

18. $x + 2y \leq 6$

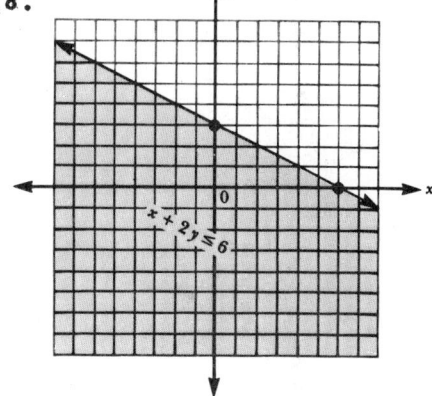

19. $\frac{1}{2}x - y > 4$

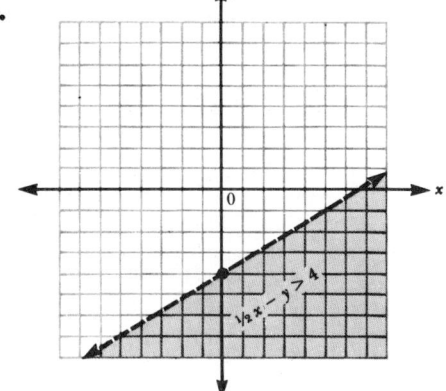

20. $x + y + 6 < 0$

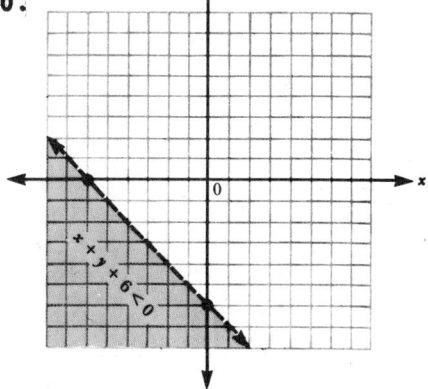

REVIEW EXERCISES (Page 418)
CHAPTER 10

1. Points plotted: A(3, 6), B(-3, 2), E(2, 0), C(5, -1), D(-3, -3), F(0, -4)

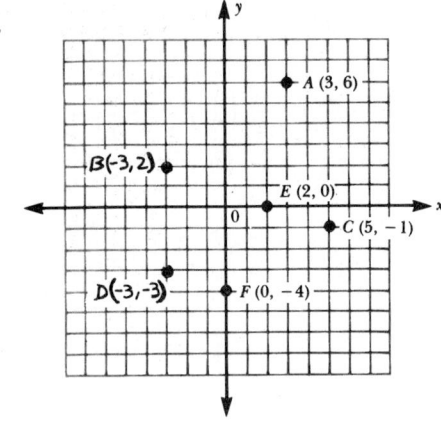

2. $x = 2$

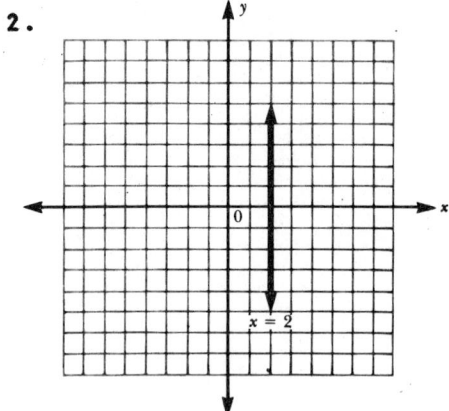

3. $y = -3$

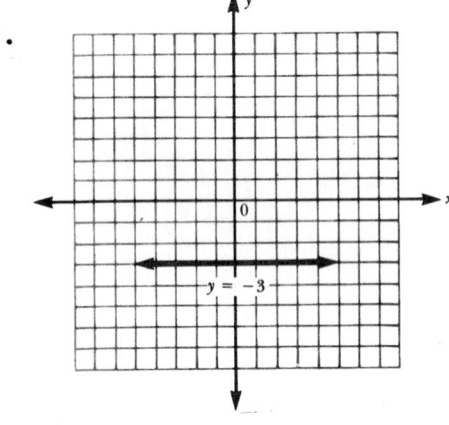

4. $y + 4 = 0$

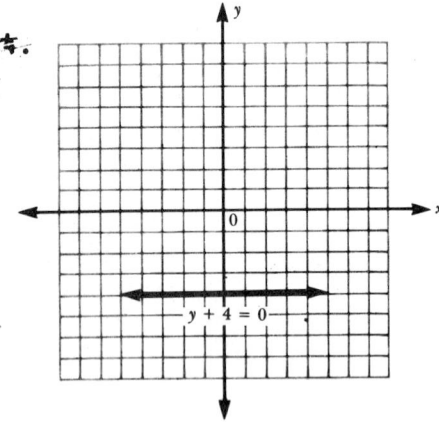

5. $x + 6 = 0$

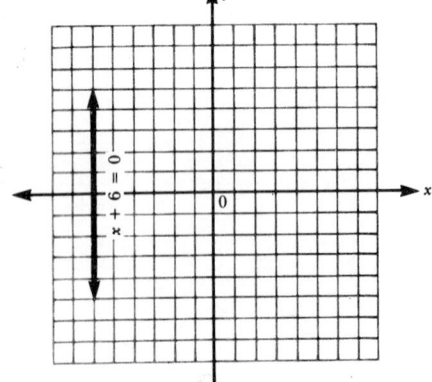

6. $y = -3x + 4$

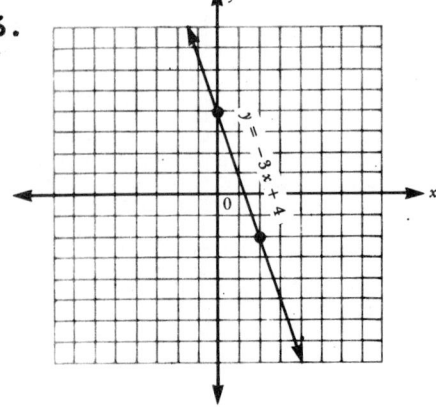

7. $y = 2x - 6$

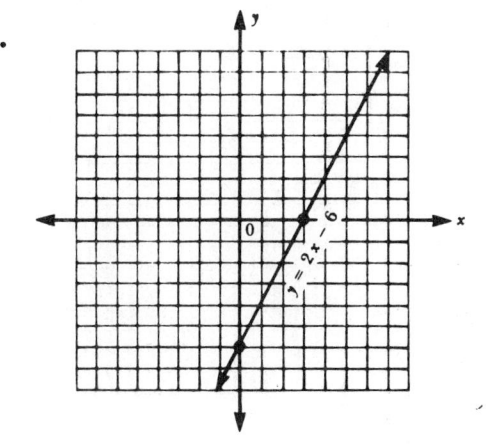

8. $y = 4x$

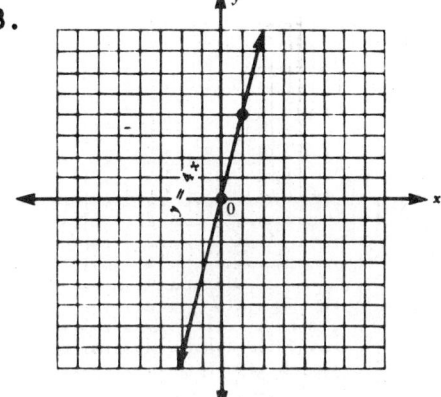

9. $y = -3x$

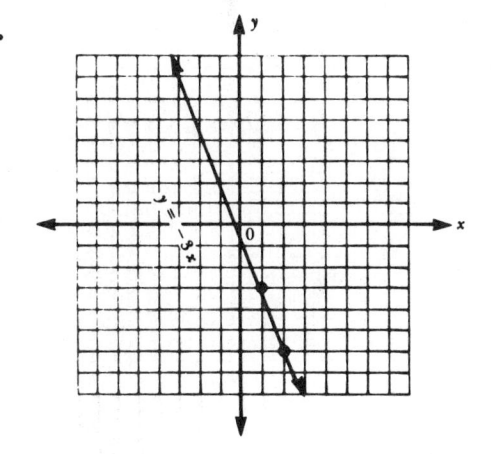

10. $y = 5x + 3$

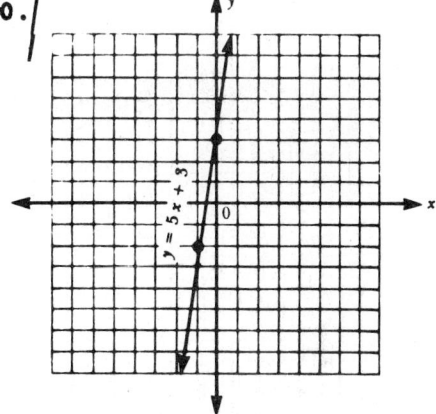

11. $y = 5x + 6$

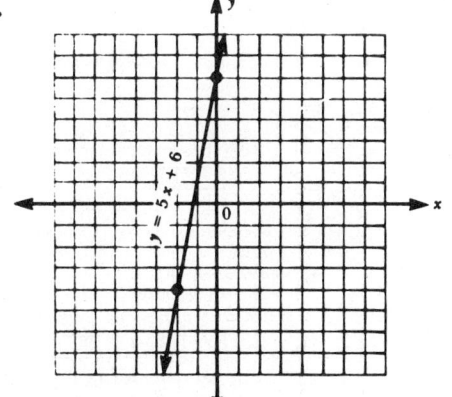

12. $3y - 2x = -6$

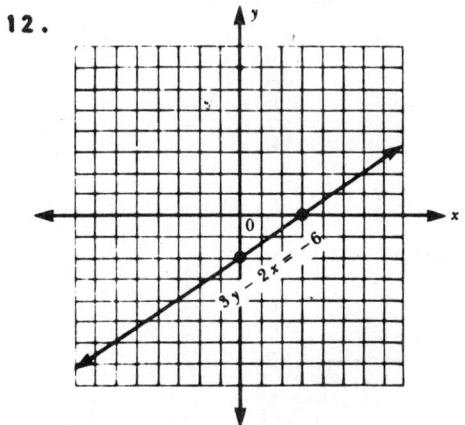

13. 2y + x = 5

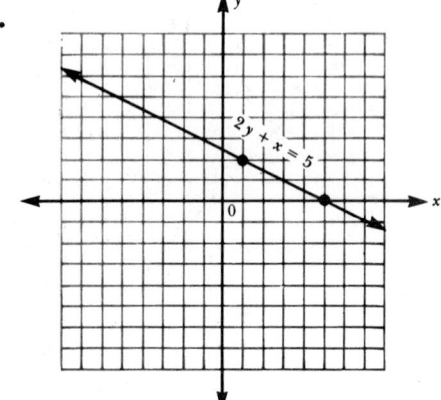

14. 3x − y + 4 = 0

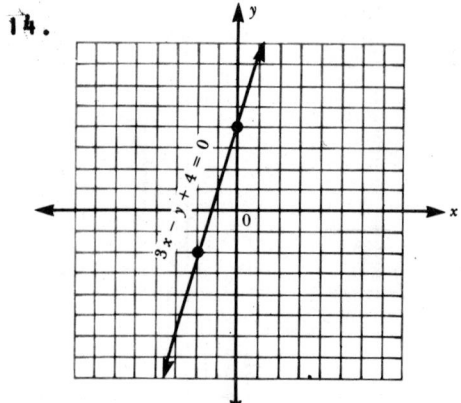

15. 5x + 3y − 6 = 0

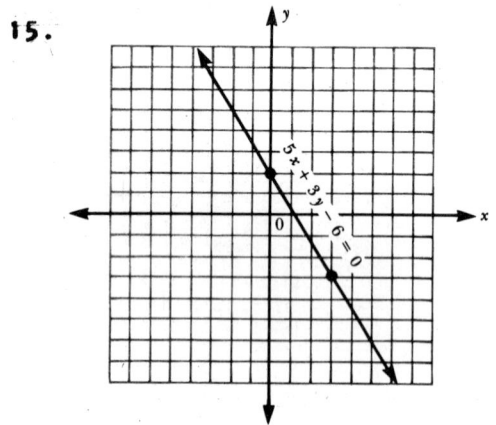

16. y = 4

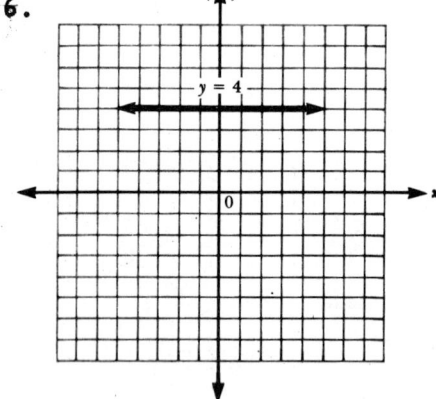

17. x = −3

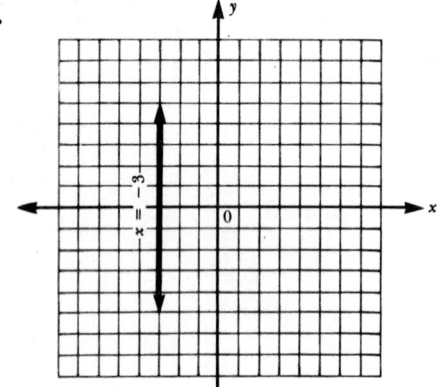

18. 2x − 5y = −24

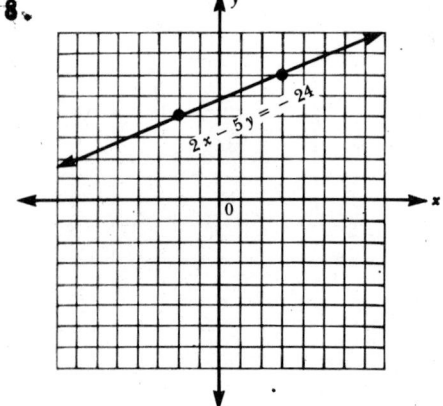

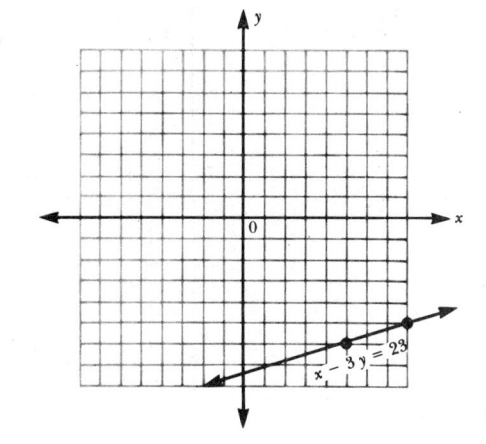

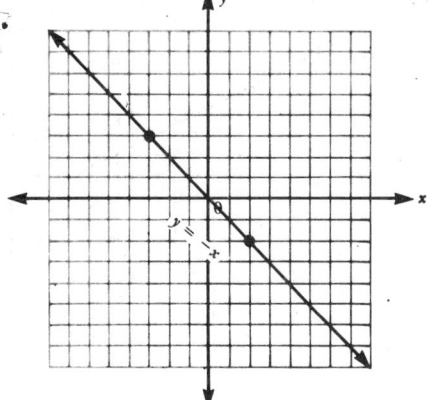

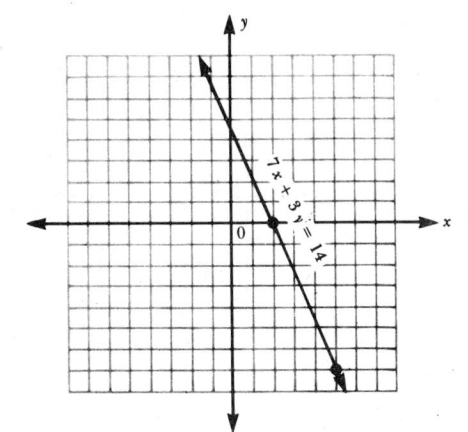

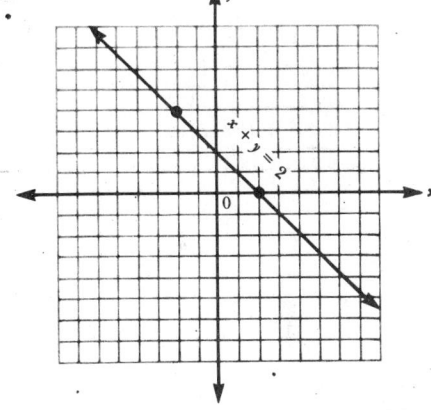

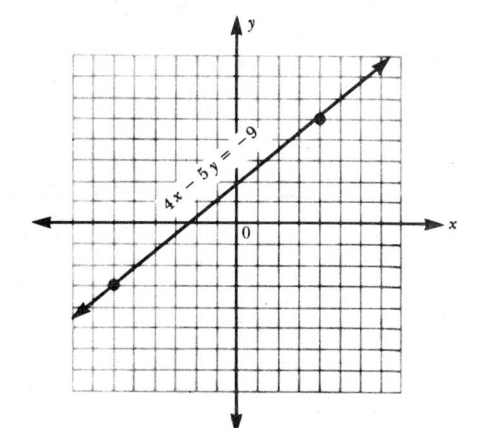

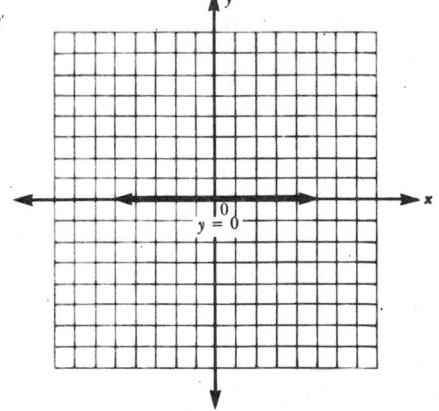

19.

20.

21.

22.

23.

24.

25. $x = 4$

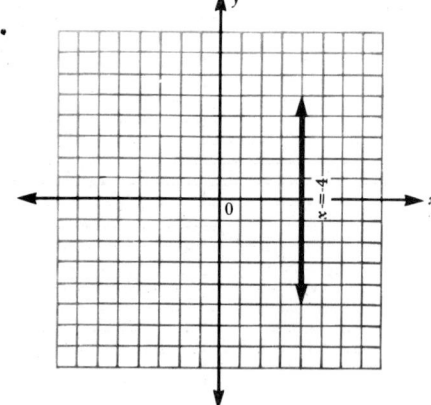

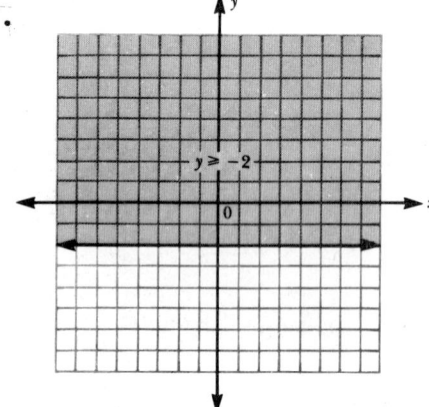

26. $x < 8$

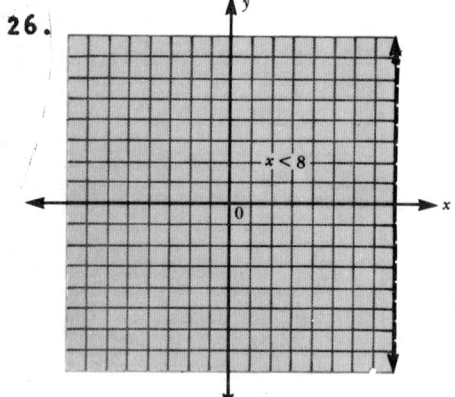

27. $y \geq -2$

28. $x + 5 < 0$

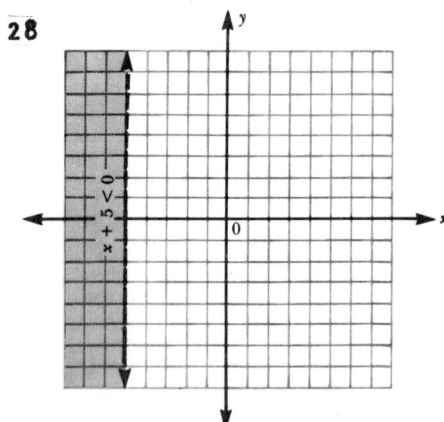

29. $y - 6 > 2$

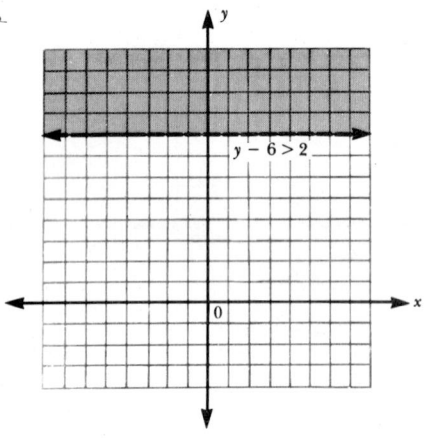

30. $x + y > 4$

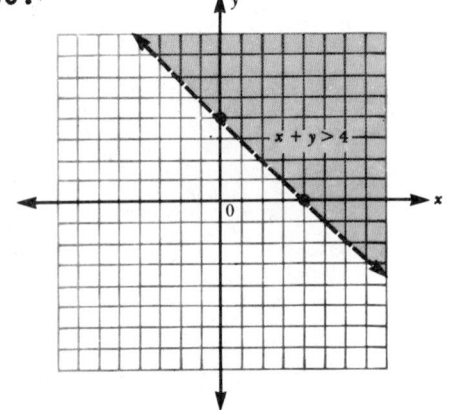

31. $-y \leq 3x$

32. $2x - 5y \geq 10$

33. $4x - 3y > 12$

34. $3x + 5y \leq 15$

35. $2x - 8 < 16 + y$

(12, 0)

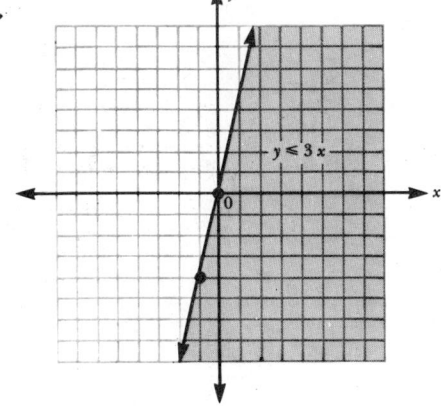

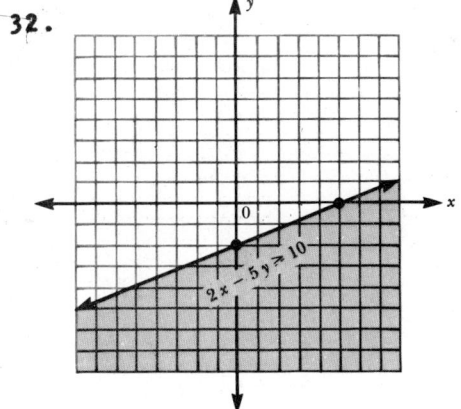

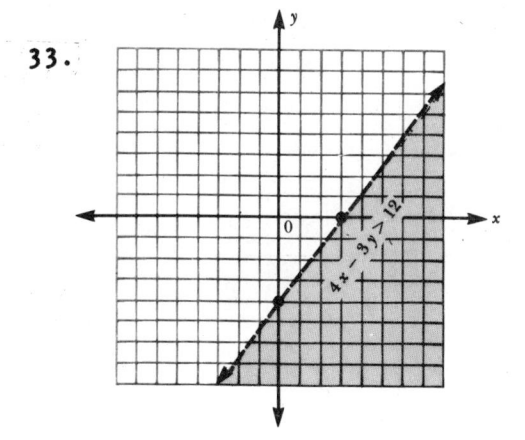

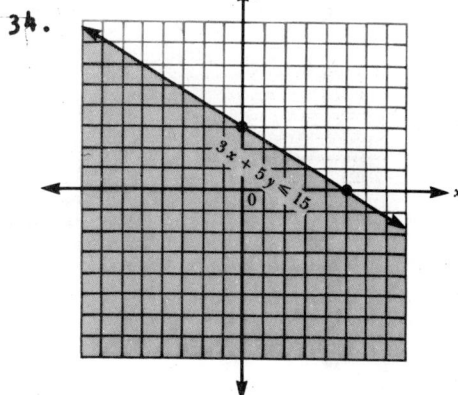

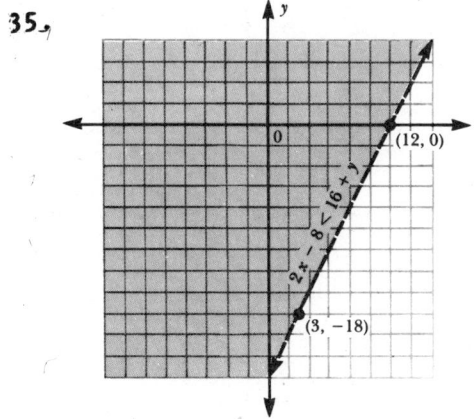

(3, −18)

(Scale of graph is in multiples of 2 for both the x and y axes.)

CHAPTER 11
EXERCISE 11-1 (Page 421)

1. No
2. Yes
3. No
4. No
5. Yes
6. Yes
7. No
8. No
9. No
10. Yes

EXERCISE 11-2 (Page 429)

1. (1, 2) Consistent

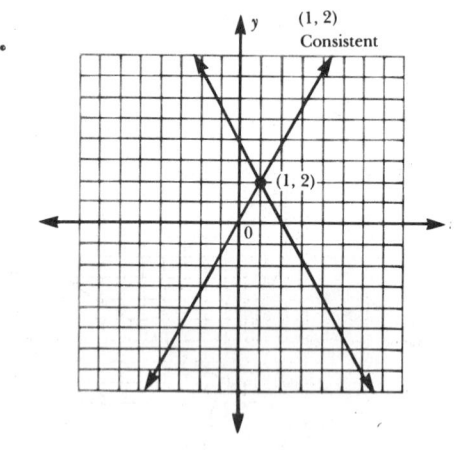

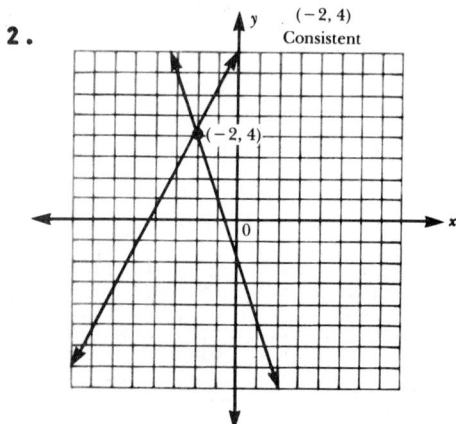

2. (−2, 4) Consistent

3. Inconsistent

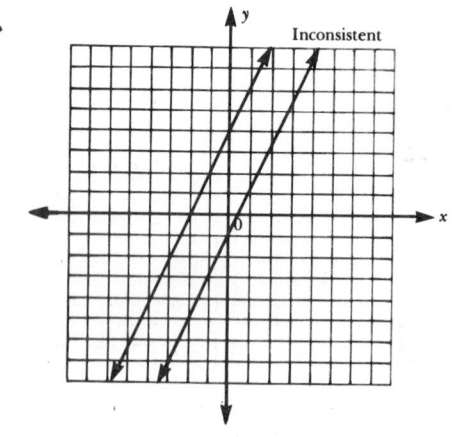

4. (0, 1) Consistent

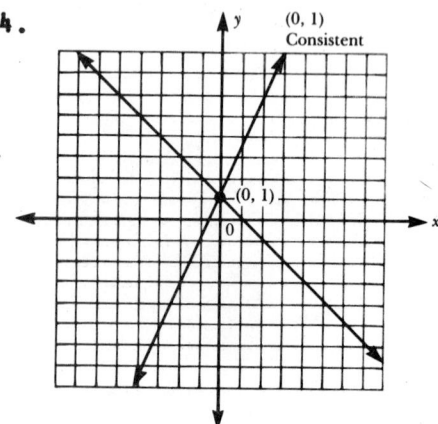

5. (2, 3) Consistent

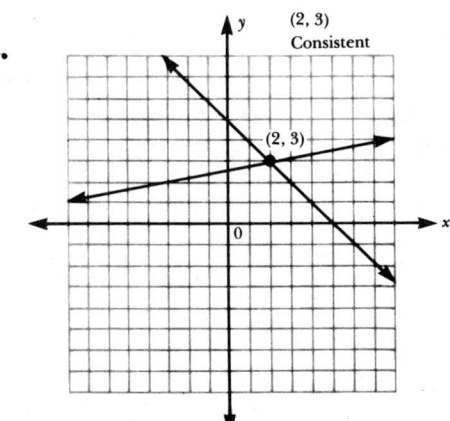

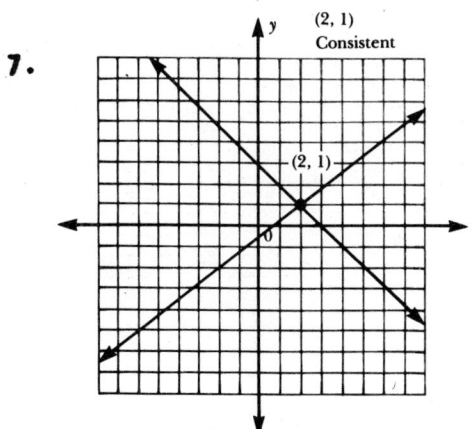

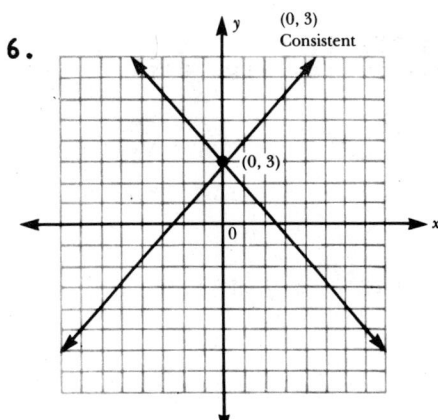

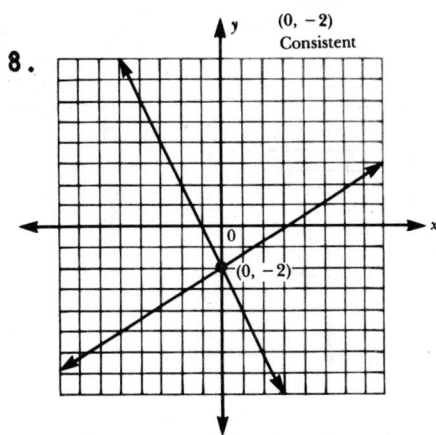

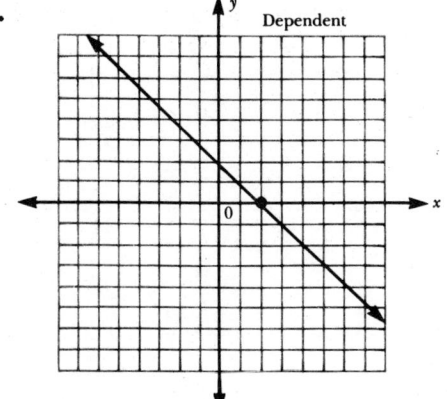

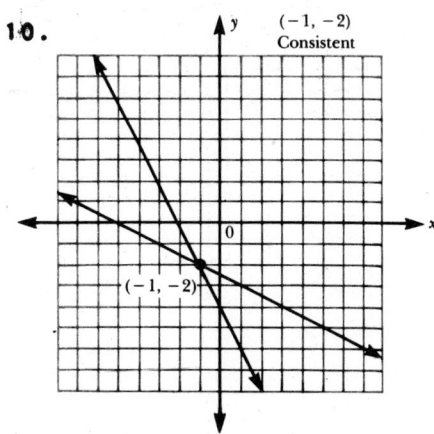

6. (0, 3) Consistent

7. (2, 1) Consistent

8. (0, −2) Consistent

9. Dependent

10. (−1, −2) Consistent

11. (2, 8) Consistent

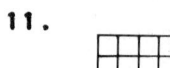

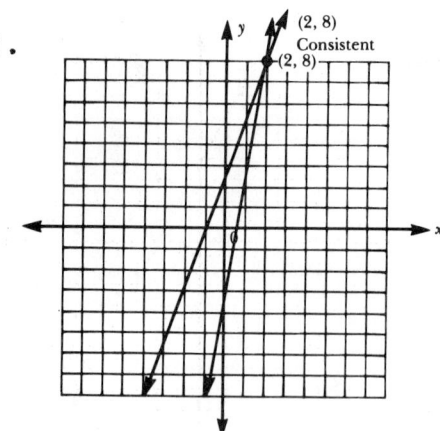

12. (0, 3) Consistent

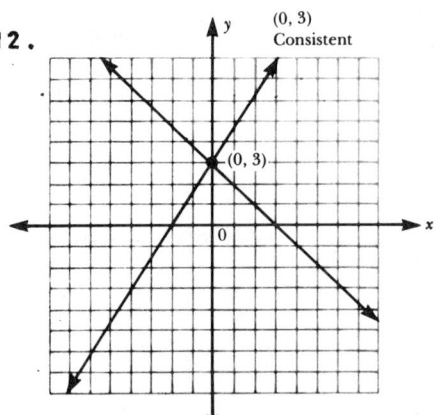

13. (1, −1) Consistent

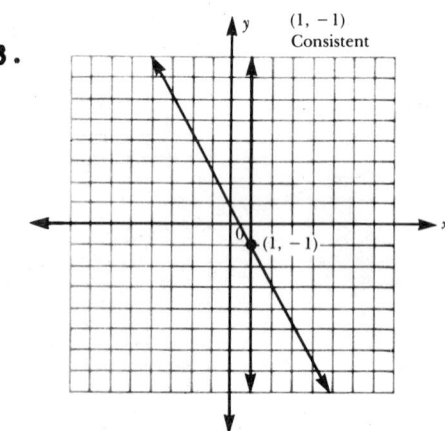

14. (3, 3) Consistent

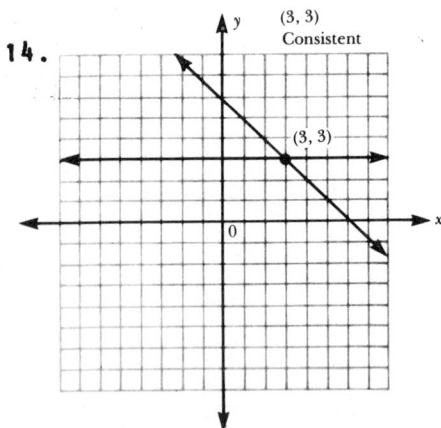

15. Inconsistent

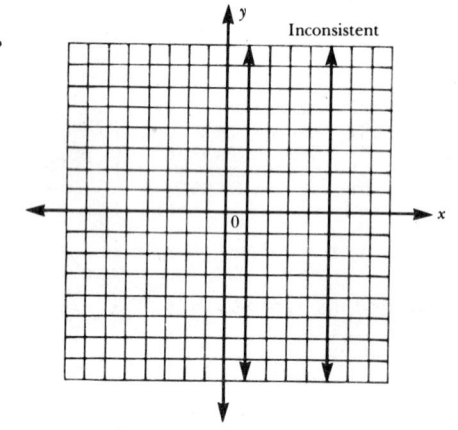

16. (−1, 2) Consistent

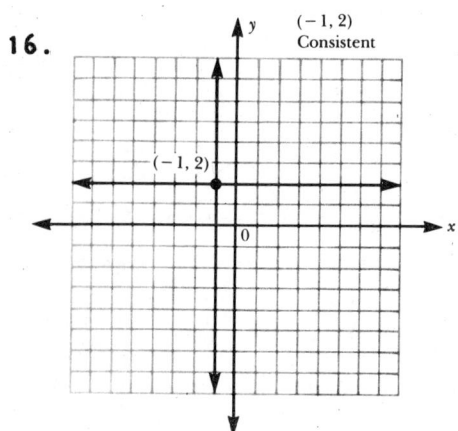

17. (0, 0) Consistent

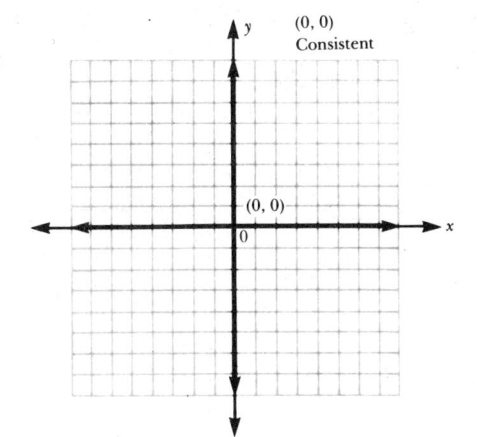

18. Dependent

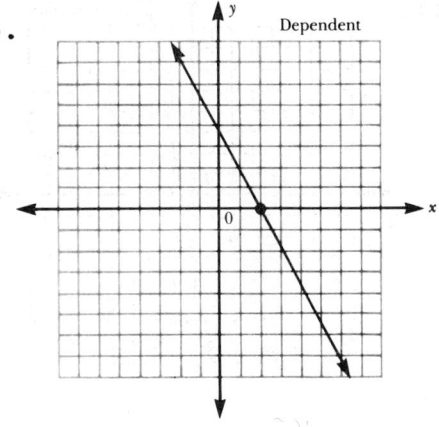

19. (−5, 3) Consistent

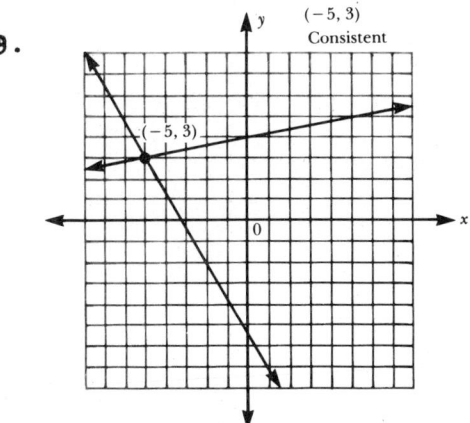

20. (7, 1) Consistent

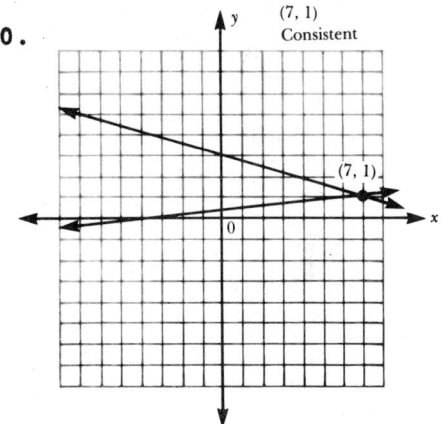

EXERCISE 11-3 (Page 437)

1. (6,2)
2. (3,1)
3. (-36,15)
4. (-8,-2)
5. (5,-3)
6. (-2,-4)
7. (1,0)
8. (-3,-5)
9. ($\frac{2}{3}$, -2)
10. ($\frac{25}{2}$, -7)
11. (1,3)
12. (9,-7)
13. (-6,-7)
14. Dependent; all points are common.
15. ($\frac{1}{2}$, $\frac{1}{2}$)
16. ($-\frac{1}{3}$, $\frac{2}{3}$)
17. Inconsistent; there are no points common.
18. ($\frac{2}{5}$, $\frac{8}{5}$)
19. ($\frac{4}{9}$, $\frac{1}{3}$)
20. ($\frac{7}{4}$, $\frac{15}{2}$)

EXERCISE 11-4 (Page 442)

1. (8,8)
2. (6,6)
3. (2,4)
4. (-2,-1)
5. (2,4)
6. (-4,-7)
7. (-6,0)
8. (1,0)
9. (-12,-18)
10. (-1,1)
11. (5,2)
12. (-9,-5)
13. (2,-4)
14. (3,5)
15. (4,0)
16. (4,3)
17. (3,-6)
18. (6,3)
19. $(\frac{7}{2}, -\frac{3}{4})$
20. $(-\frac{1}{2}, -\frac{3}{2})$

EXERCISE 11-5A (Page 446)

1. 48; 25
2. 29; 17
3. 20; 19
4. 22; 17
5. 20, 19
6. 10; 7
7. 20; 93
8. 129; 171
9. 8; 1
10. 37; 13

EXERCISE 11-5C (Page 449)

1. 230 nickels
   115 dimes
2. 87 dimes
   99 quarters
3. 17 nickels
   148 dimes
4. 4 dimes
   28 quarters
5. 27 five dollar chips
   9 ten dollar chips

EXERCISE 11-5D (Page 451)

1. $27\frac{1}{2}$ pounds waxed beans
   $22\frac{1}{2}$ pounds kidney beans
2. 60 pounds Fave's
   40 pounds Cecci's
3. 4 tons of #1 cr
   8 tons of #2
4. 14 pounds A1
   21 pounds A2
5. 9 pounds McIntosh
   6 pounds Cortland

EXERCISE 11-5E (Page 453)

1. $800 at 8%
   $1700 at 11%
2. $700 at 5%
   $4,300 at 7%
3. $7,600 at 4%
   $5,000 at $5\frac{1}{2}$%
4. $30,000 at 6%
   $18,000 at 10%
5. $2,500

EXERCISE 11-5F (Page 456)

1. 2 mph
2. 3 mph
3. 50 mph
4. still air: 450 mph
   speed of wind: 150 mph
5. motorboat: $10\frac{1}{2}$ mph
   current: $4\frac{1}{2}$ mph

REVIEW EXERCISES (Page 457)
CHAPTER 11

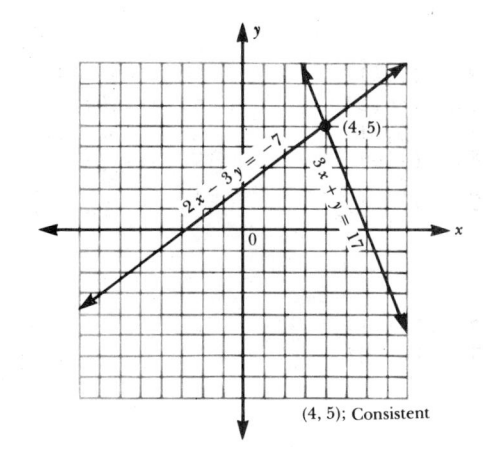

1. (4, 5); Consistent

2. (−3, −1); Consistent

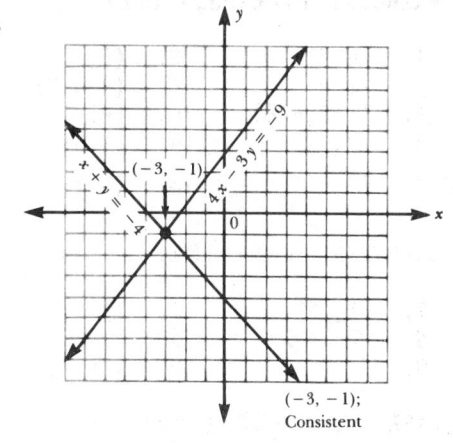

3. Inconsistent

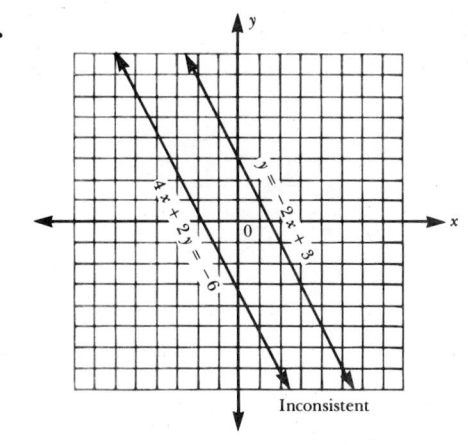

4. Dependent

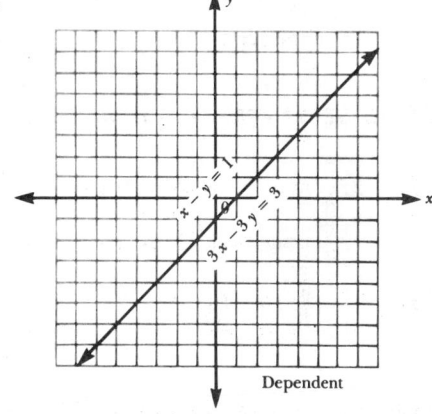

5. (0, 0); Consistent

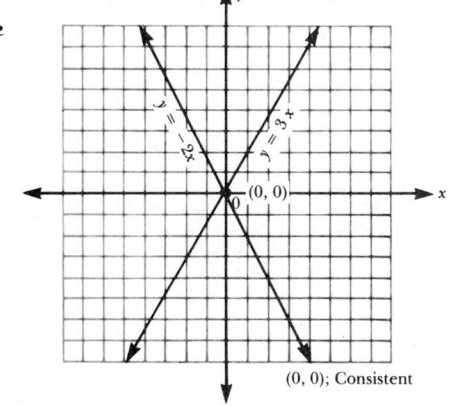

6. Inconsistent

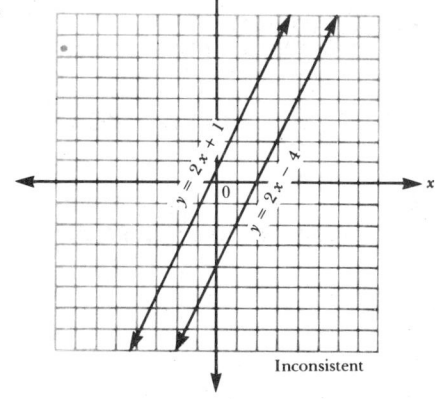

7. (4, −6); Consistent

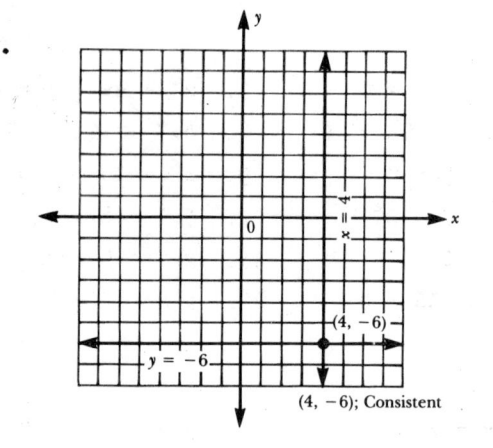

8. (−3/2, −3/2); Consistent

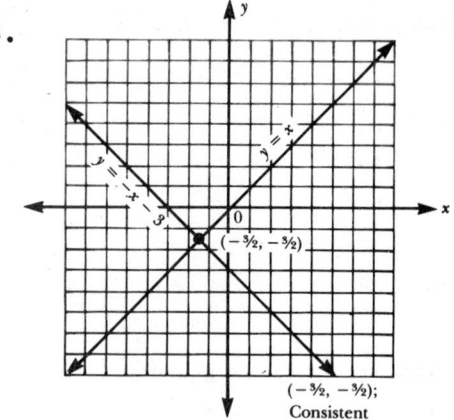

9. Dependent

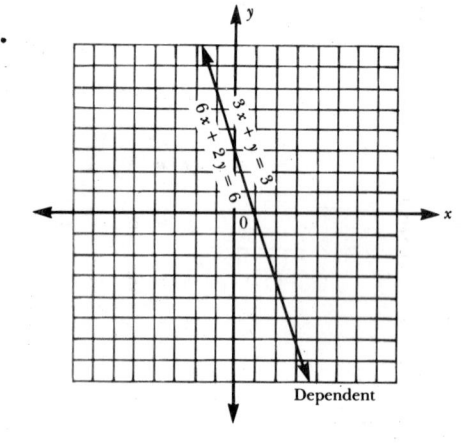

10. (5/3, 5/3); Consistent

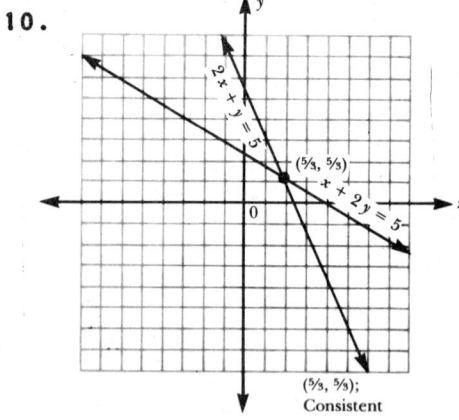

11. (4,4)
12. (−4,14)
13. (−2,0)
14. (½, 5)
15. (−1,−2)
16. (5/3, 1)
17. (9, −8)
18. (−2/3, −1/3)
19. (5/11, −7/11)
20. (−2, −2)
21. 61; 19
22. 48; 13
23. Boat: 18 mph
    Current: 6 mph
24. $5,625 at 8%
    $4,375 at 12%
25. 17½ pounds of first brand
    12½ pounds of second brand

CHAPTER 12
EXERCISE 12-1 (Page 462)

1. $x^2 + 6x - 40 = 0$
2. $2x^2 - 8x - 11 = 0$
3. $x^2 + 4x - 5 = 0$
4. $x^2 - 5x - 20 = 0$
5. $2x^2 + 16x - 66 = 0$
6. $4x^2 - 12x - 27 = 0$
7. $2x^2 - 6x + 3 = 0$
8. $x^2 - 6x + 5 = 0$
9. $x^2 - 7x + 2 = 0$
10. $2x^2 - x + 12 = 0$
11. $x^2 - 5x - 7 = 0$
12. $5x^2 + 8x + 2 = 0$
13. $x^2 - 4 = 0$
14. $8x^2 + 5 = 0$
15. $4x - 36 = 0$
16. $x^2 + 10x + 18 = 0$
17. $x^2 - 13x + 36 = 0$
18. $x^2 + 4x - 5 = 0$

EXERCISE 12-2 (Page 467)

1. $x = -1, x = 7$
2. $x = 2, x = 7$
3. $x = -2, x = 5$
4. $x = 2, x = \frac{3}{2}$
5. $x = 5, x = -5$
6. $x = +4, x = -4$
7. $x = 3, x = -3$
8. $x = 5, x = -5$
9. $x = 0, x = 9$
10. $x = 0, x = -8$
11. $x = 0, x = 4$
12. $x = 0, x = -\frac{5}{2}$
13. $x = 3, x = -\frac{5}{3}$
14. $x = 1, x = -10$
15. $x = -2, x = 5$
16. $x = -1, x = \frac{14}{3}$
17. $x = 3, x = 3$
18. $x = 3, x = -\frac{1}{2}$
19. $x = -3, x = 9$
20. $x = 4, x = -\frac{3}{2}$
21. $x = 6, x = -6$
22. $x = 3, x = 9$
23. $x = 2, x = -\frac{5}{4}$
24. $x = 8, x = -5$
25. $x = 0, x =$
26. $x = +3, x = -3$
27. $x = 0, x = 3$
28. $x = 9$
29. $x = 0$
30. $x = 2$

EXERCISE 12-3 (Page 471)

1. $x = 3, x = -3$
2. $x = 5, x = -5$
3. $x = 3, x = -3$
4. $x = 5, x = -5$
5. $x = 1, x = -1$
6. $x = 10, x = -10$
7. $x = 3, x = -3$
8. $x = 3, x = -3$
9. $x = 2\sqrt{3}, x = -2\sqrt{3}$
10. $x = \sqrt{2}, x = -\sqrt{2}$
11. $x = 2\sqrt{2}, x = -2\sqrt{2}$
12. $x = 3\sqrt{3}, x = -3\sqrt{3}$
13. $x = 2\sqrt{7}, x = -2\sqrt{7}$
14. $x = \sqrt{2}, x = -\sqrt{2}$
15. $x = 4\sqrt{2}, x = -4\sqrt{2}$
16. $x = \sqrt{10}, x = -\sqrt{10}$

EXERCISE 12-4 (Page 476)

1. $x = -1, x = 3$
2. $x = 5, x = 5$
3. $x = 1, x = -4$
4. $x = 2, x = -7$
5. $x = 1, x = 4$
6. $x = 0, x = -3$
7. $x = -2, x = -\frac{5}{2}$
8. $x = 4, x = -\frac{9}{2}$
9. $x = -1 + \sqrt{10}, x = -1 - \sqrt{10}$
10. $x = 3 + 2\sqrt{2}, x = 3 - 2\sqrt{2}$
11. $x = 3 + \sqrt{17}, x = 3 - \sqrt{17}$
12. $x = 12 + \sqrt{154}, x = 12 - \sqrt{154}$

281

EXERCISE 12-4 (Cont'd)

13. $x = 3 + 2\sqrt{3}, x = 3 - 2\sqrt{3}$
14. $x = 2 + \sqrt{10}, x = 2 - \sqrt{10}$
15. $x = 2 + \sqrt{2}, x = 2 - \sqrt{2}$
16. $x = 1 + \frac{\sqrt{15}}{3}, x = 1 - \frac{\sqrt{15}}{3}$
17. $x = \frac{3 + \sqrt{15}}{2}, x = \frac{3 - \sqrt{15}}{2}$
18. $x = \frac{-1 + \sqrt{6}}{2}, x = \frac{-1 - \sqrt{6}}{2}$
19. $x = -\frac{1}{5}, x = -1$
20. $x = \frac{-1 + \sqrt{22}}{3}, x = \frac{-1 - \sqrt{22}}{3}$

EXERCISE 12-5 (Page 481)

1. $x = -3, x = 5$
2. $x = 1, x = 7$
3. $x = -5, x = 7$
4. $x = 2, x = -4$
5. $x = -1, x = -\frac{2}{3}$
6. $x = -2, x = -\frac{1}{2}$
7. $x = -2, x = \frac{1}{4}$
8. $x = 3, x = \frac{1}{3}$
9. $x = 2, x = -2$
10. $x = 0, x = 6$
11. $x = \sqrt{3}, x = -\sqrt{3}$
12. $x = 0, x = 0$
13. $x = -3 + \sqrt{6}, x = -3 - \sqrt{6}$
14. $x = 5 + \sqrt{30}, x = 5 - \sqrt{30}$
15. $x = -1 + \sqrt{2}, x = -1 - \sqrt{2}$
16. $x = 2 + \sqrt{3}, x = 2 - \sqrt{3}$
17. $x = \frac{13 + \sqrt{137}}{4}, x = \frac{13 - \sqrt{137}}{4}$
18. $x = -1, x = -\frac{3}{5}$
19. $x = \frac{-1 + \sqrt{33}}{4}, x = \frac{-1 - \sqrt{33}}{4}$
20. $x = \frac{\sqrt{51}}{3}, x = \frac{-\sqrt{51}}{3}$
21. $x = \frac{4 + \sqrt{6}}{5}, x = \frac{4 - \sqrt{6}}{5}$
22. $x = -6 + 3\sqrt{3}, x = -6 - 3\sqrt{3}$
23. $x = \frac{3 + \sqrt{6}}{2}, x = \frac{3 - \sqrt{6}}{2}$
24. $x = \frac{9 + \sqrt{93}}{6}, x = \frac{9 - \sqrt{93}}{6}$
25. $x = \frac{7 + \sqrt{41}}{2}, x = \frac{7 - \sqrt{41}}{2}$
26. $x = \frac{3 + \sqrt{17}}{4}, x = \frac{3 - \sqrt{17}}{4}$
27. $x = 1 + \sqrt{3}, x = 1 - \sqrt{3}$
28. $x = 2 + \sqrt{6}, x = 2 - \sqrt{6}$
29. $x = 3, x = \frac{1}{3}$
30. $x = \frac{5 + \sqrt{21}}{2}, x = \frac{5 - \sqrt{21}}{2}$
31. $x = 3.8, x = -0.8$
32. $x = 0.4, x = -7.4$
33. $x = 1.6, x = -.6$
34. $x = 1.3, x = 0.2$
35. $x = 0.4, x = -1.7$
36. $x = 1.3, x = -0.4$

SUPPLEMENTARY EXERCISE 12-5 (Page 483)

1. $x = 0, x = 6$
2. $x = 5, x = -5$
3. $x = 3, x = -7$
4. $x = 6, x = -6$
5. $x = \frac{-5 + \sqrt{89}}{4}, x = \frac{-5 - \sqrt{89}}{4}$
6. $x = 3 + \sqrt{2}, x = 3 - \sqrt{2}$
7. $x = 0, x = -3$
8. $x = 2\sqrt{2}, x = -2\sqrt{2}$
9. $x = \frac{11 + \sqrt{21}}{10}, x = \frac{11 - \sqrt{21}}{10}$

SUPPLEMENTARY EXERCISE 12-5 (Cont'd)

10. $x = \frac{3\sqrt{2}}{2}$, $x = \frac{-3\sqrt{2}}{2}$ 11. $x = 4$, $x = -\frac{7}{2}$ 12. $x = \frac{1}{2}$, $x = \frac{3}{4}$

13. $x = \frac{3 + \sqrt{7}}{2}$, $x = \frac{3 - \sqrt{7}}{2}$ 14. $x = 3$, $x = \frac{1}{3}$ 15. $x = 4$ 16. $x = 6, -7$

EXERCISE 12-6 (Page 490)

1. $y = x^2$

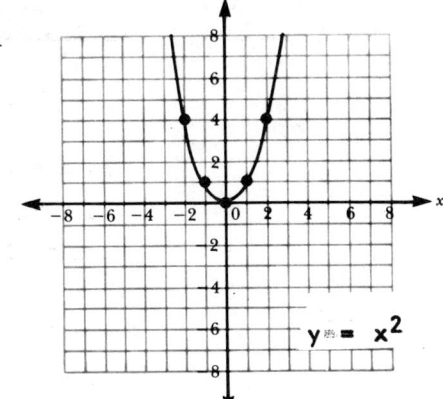

2. $y = 4x^2$

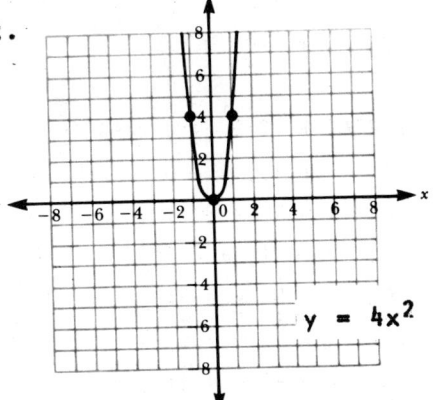

3. $y = 3x^2$

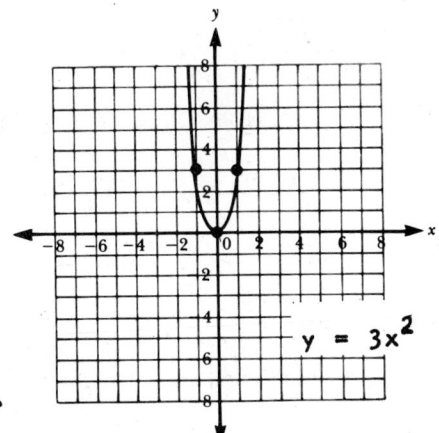

4. $y = x^2 + 1$

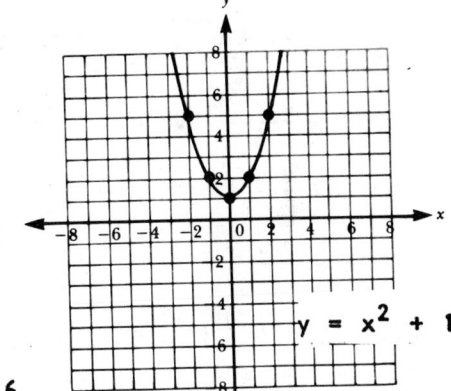

5. $y = x^2 - 1$

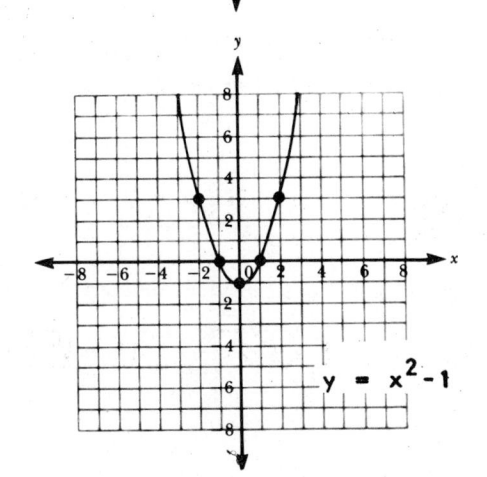

6. $y = x^2 + 4$

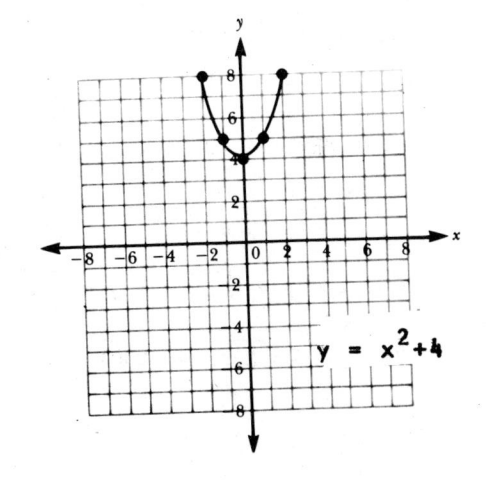

EXERCISE 12-6 (Cont'd)

7. $y = x^2 - 2x$

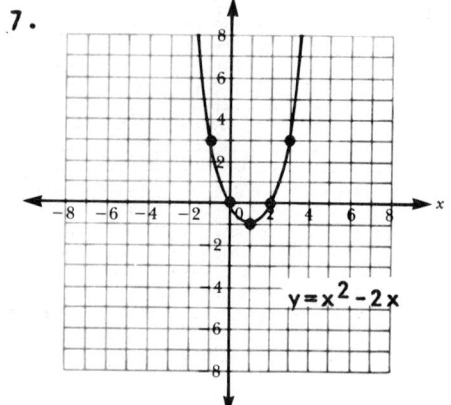

8. $y = x^2 + 6x$

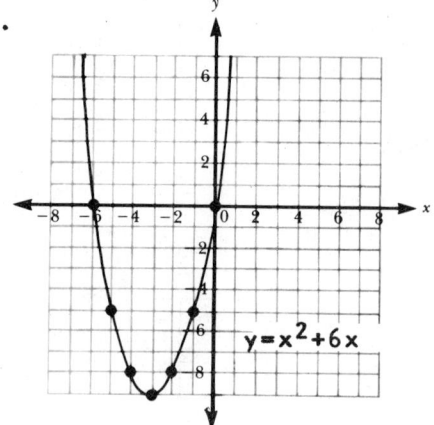

9. $y = 2x^2 - x$

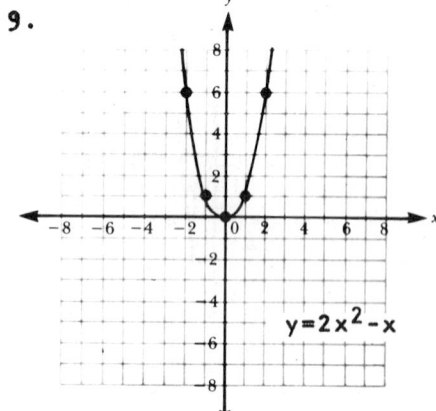

10. $y = 3x^2 - 4x$

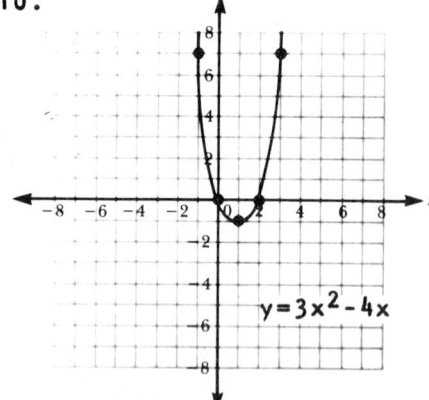

11. $y = x^2 - 2x + 2$

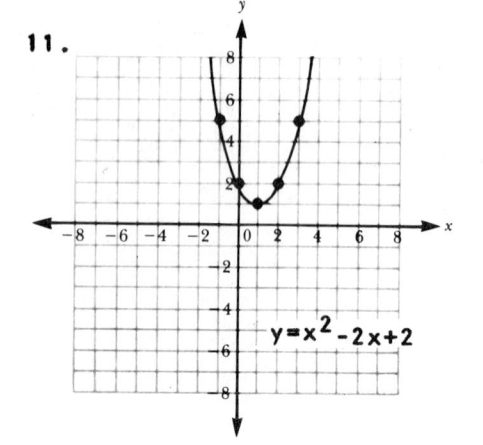

12. $y = 2x^2 - 3x + 2$

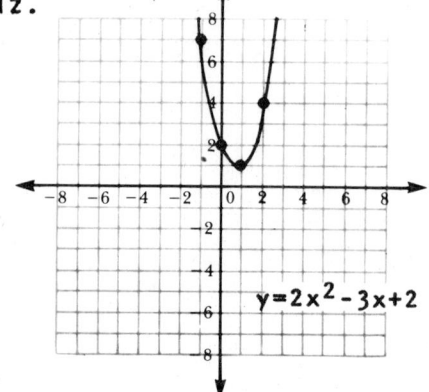

EXERCISE 12-6 (Cont'd)

13. $y = x^2 - 6x + 8$

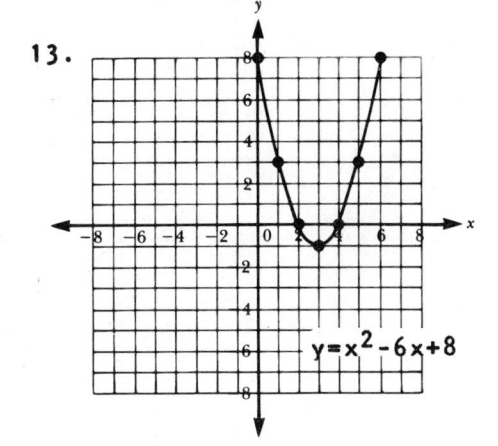

14. $y = x^2 - 2x - 3$

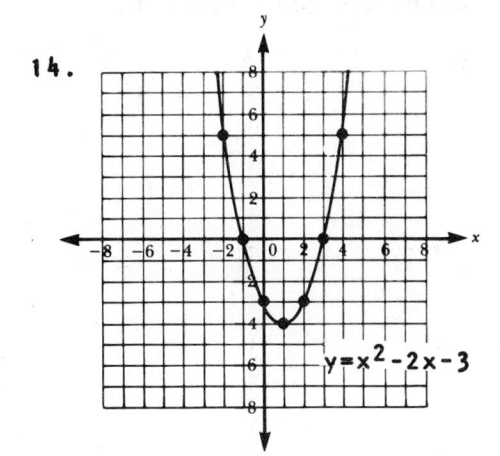

15. $y = 2x^2 + 4x - 5$

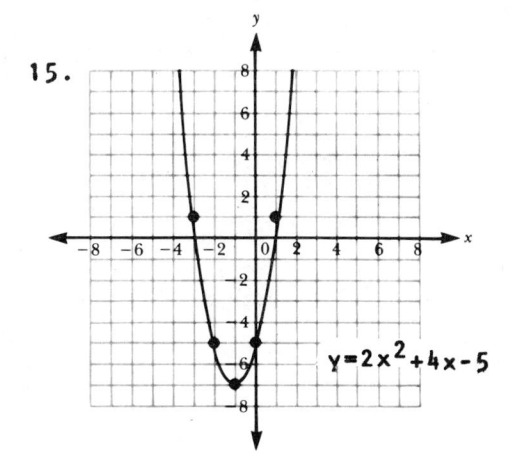

16. $y = 3x^2 - x - 6$

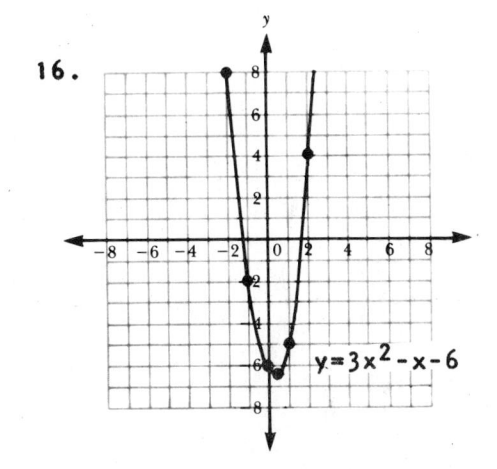

EXERCISE 12-7 (Page 494)

1. $x^2 - 25 = 0$ −5, +5

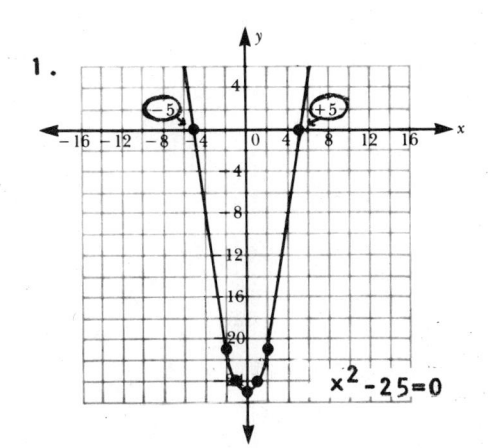

2. $x^2 - 12 = 0$ −3.5, 3.5

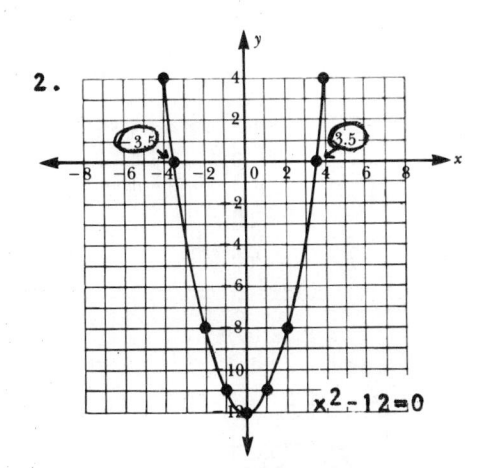

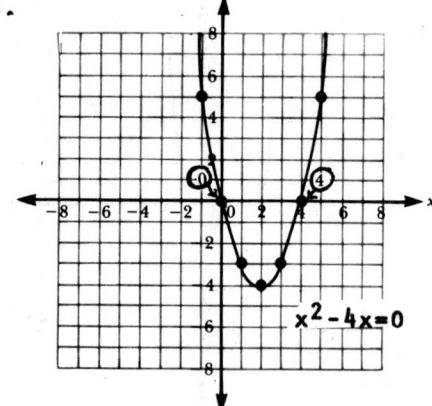

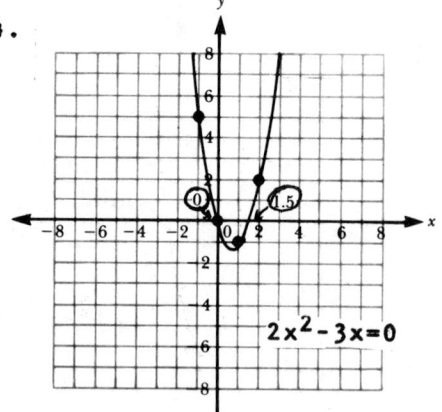

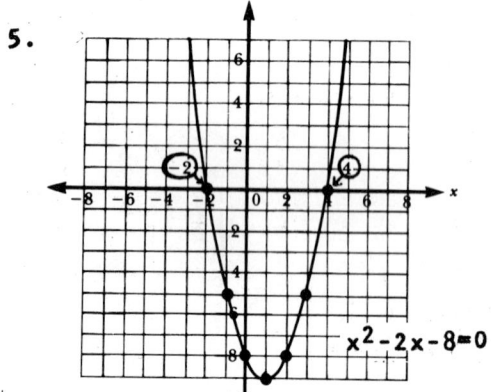

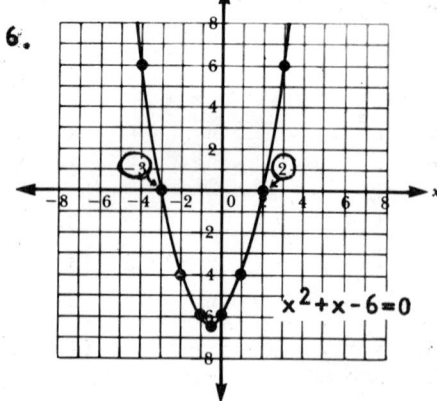

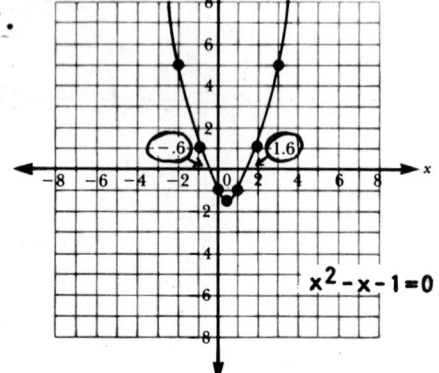

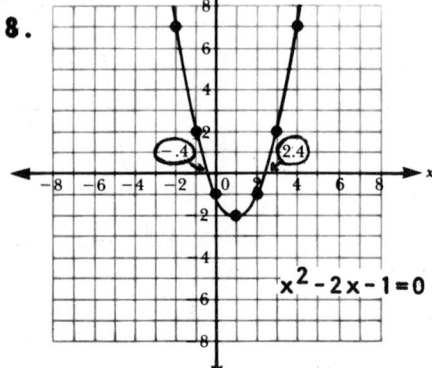

EXERCISE 12-7 (Cont'd)

3. $x^2 - 4x = 0$　　roots: 0, 4

4. $2x^2 - 3x = 0$　　roots: 0, 1.5

5. $x^2 - 2x - 8 = 0$　　roots: -2, 4

6. $x^2 + x - 6 = 0$　　roots: -3, 2

7. $x^2 - x - 1 = 0$　　roots: -.6, 1.6

8. $x^2 - 2x - 1 = 0$　　roots: -.4, 2.4

EXERCISE 12-7 (Cont'd)

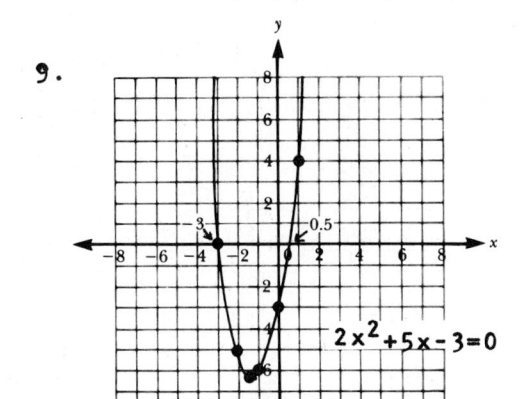

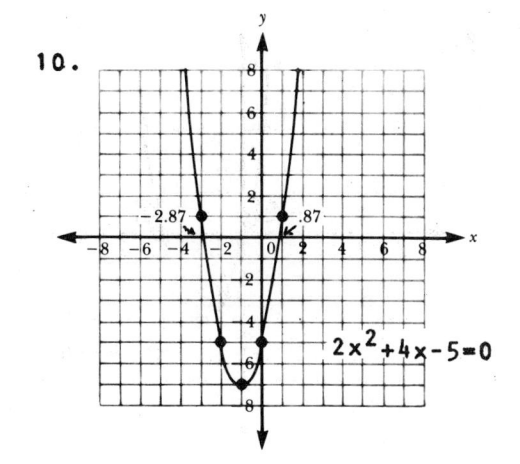

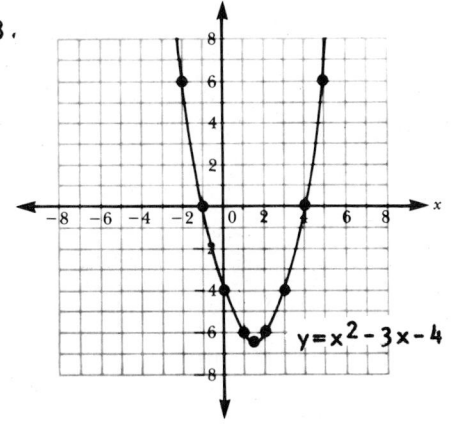

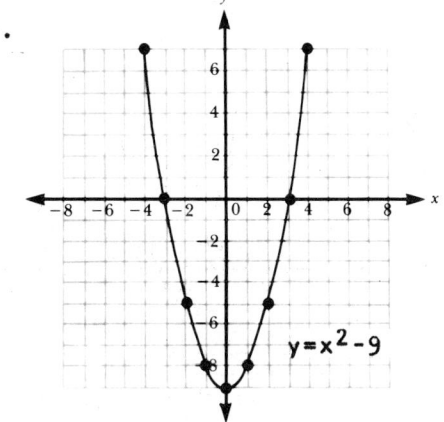

9. $2x^2+5x-3=0$

10. $2x^2+4x-5=0$

REVIEW EXERCISES (Page 494)
CHAPTER 12

1. $2x^2 - 5x - 6 = 0$
2. $x^2 + 6x - 4 = 0$
3. $x^2 - 3x + 2 = 0$
4. $x = 0, x = 3$
5. $x = 4, x = -4$
6. $x = -2, x = -4$
7. $x = 2, x = 7$
8. $x = 2, x = -6$
9. $x = -4, x = -4$
10. $x = -3, x = 7$
11. $x = 5, x = -2$
12. $x = 0, x = 0$
13. $x = 4, x = -4$
14. $x = 2\sqrt{2}, x = -2\sqrt{2}$
15. $x = 3\sqrt{2}, x = -3\sqrt{2}$
16. $x = -1, x = -3$
17. $x = 1, x = 2$
18. $x = \frac{1}{2}, x = -3$
19. $x = -3 + \sqrt{10}, x = -3 - \sqrt{10}$
20. $x = 2, x = 3$
21. $x = -2 + \sqrt{6}, x = -2 - \sqrt{6}$
22. $x = .9, x = -10.9$

23. $y = x^2 - 3x - 4$

24. $y = x^2 - 9$

REVIEW EXERCISES
CHAPTER 12 (Cont'd)

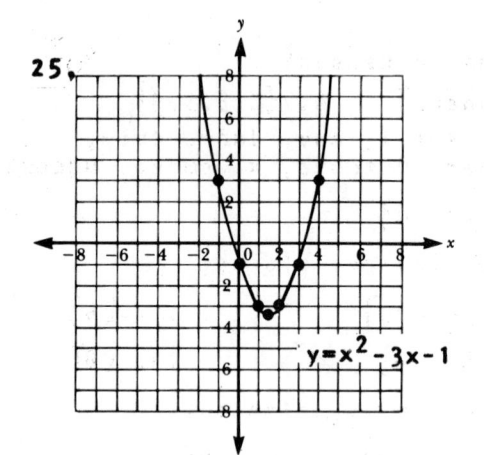

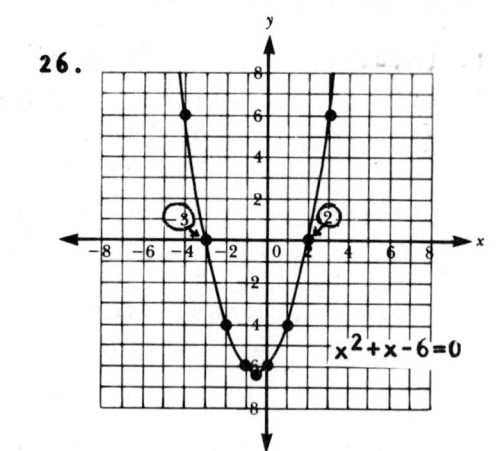

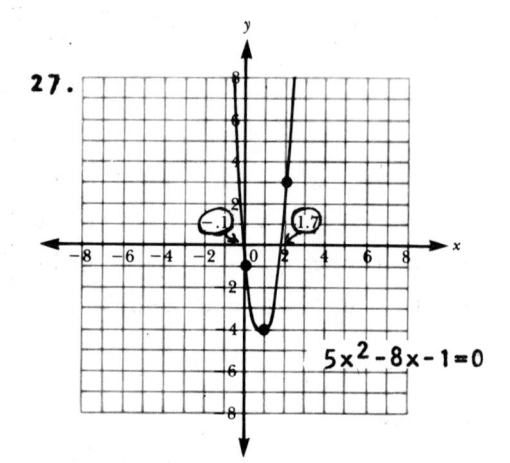

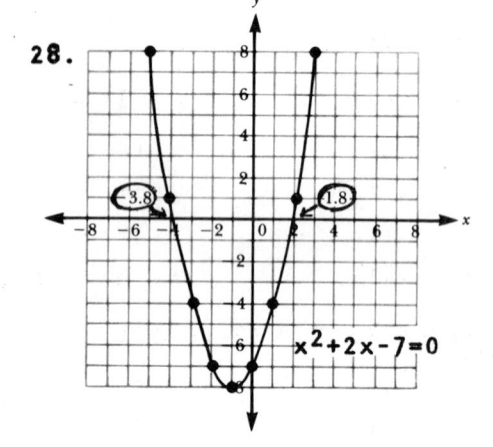

25. $y = x^2 - 3x - 1$

26. $x^2 + x - 6 = 0$ -3, 2

27. $5x^2 - 8x - 1 = 0$ -.1, 1.7

28. $x^2 + 2x - 7 = 0$ -3.8, 1.8

ANSWERS TO APPENDIX A

EXERCISE A-1 (Page 500)

1. {m,u,s,k}
2. {Alabama, Arizona, Arkansas}
3. {Lincoln, Garfield, McKinley, Kennedy}
4. {2,3,5,8}
5. {1,3,9}
6. {January, March, April, May, June, July, August, September, October, November, December}
7. {Albany}
8. {first, second, third}
9. {red, orange, yellow, green, blue, indigo, violet}
10. {Datsun, Honda, Isuzu, Subreau, Toyota}
11. {4}
12. {15}
13. {0}
14. {2}
15. {8}
16. {x/x is a lowercase letter of the English Alphabet}
17. {x/x is a multiple of 5 less than or equal to 100}
18. {x/x is a month of the year with only 30 days}
19. {x/x is an ocean}
20. {x/x is one of the three R's}
21. True
22. False
23. True
24. False
25. False
26. False
27. True
28. True
29. False
30. True

EXERCISE A-2 (Page 503)

1. Finite
2. Infinite
3. Infinite
4. Finite
5. Empty
6. Empty
7. Empty
8. Finite
9. Infinite
10. Finite
11. Equivalent
12. Equivalent
13. Neither
14. Neither
15. Equivalent and equal

EXERCISE A-3 (Page 506)

1. False
2. True
3. True
4. True
5. True
6. True
7. True
8. False
9. False
10. False
11. True
12. True
13. False
14. False
15. True
16. False
17. True
18. False
19. False
20. False
21. ∅, {2}, {8}, {9}, {2,8}, {2,9}, {8,9}, {2,8,9}
22. ∅, {a}, {b}, {c}, {d}, {a,b}, {a,c}, {a,d}, {b,c}, {b,d}, {c,d}, {a,b,c}, {a,b,d}, {a,c,d}, {b,c,d}, {a,b,c,d}

EXERCISE A-4 (Page 509)

1. {1}
2. {4,8}
3. {3,7}
4. { }
5. {1,2,3,4,5,6,7,8}
6. {1,3,4,5,6,7,8,9,10}

EXERCISE A-4 (Cont'd)

7. $\{3,5,6,7,9,10\}$ 8. $\{1,2,3,5,7,9,10\}$
9. $\{4,5,6,8\}$ 10. $\{1,2\}$ 11. $\{1,2,3,5,6,7,9,10\}$
12. $\{1,2,3,4,5,6,7,8,9,10\}$ or U 13. $\{\}$ or $\emptyset$
14. $\{3,5,6,7,9,10\}$ or $\bar{A}$ 15. $\{1,4,8\}$ 16. $\{1,2,4,8\}$
17. $\{1,3,4,5,6,7,8,9,10\}$ 18. $\{1,3,7\}$ 19. $\{5,6\}$
20. $\{1,2,5,9,10\}$ 21. $P \cap Q = \{3,7\}$;
$P \cup Q = \{1,2,3,4,5,7,8,9\}$

22. $R \cap S$ ⟵●─○⟶
         4 8

$R \cup S$ = (All real numbers) ⟵⟶